ROUTLEDGE LIBRARY EDITIONS:
EVOLUTION

Volume 13

GOVERNMENT BY NATURAL SELECTION

GOVERNMENT BY NATURAL SELECTION

HUGH TAYLOR

Routledge
Taylor & Francis Group
LONDON AND NEW YORK

First published in 1915 by Methuen & Co. Ltd.

This edition first published in 2020
by Routledge
2 Park Square, Milton Park, Abingdon, Oxon OX14 4RN

and by Routledge
52 Vanderbilt Avenue, New York, NY 10017

Routledge is an imprint of the Taylor & Francis Group, an informa business

British Library Cataloguing in Publication Data
A catalogue record for this book is available from the British Library

ISBN: 978-0-367-27938-7 (Set)
ISBN: 978-0-429-31628-9 (Set) (ebk)
ISBN: 978-0-367-27251-7 (Volume 13) (hbk)
ISBN: 978-0-429-29579-9 (Volume 13) (ebk)

Publisher's Note
The publisher has gone to great lengths to ensure the quality of this reprint but points out that some imperfections in the original copies may be apparent.

Disclaimer
The publisher has made every effort to trace copyright holders and would welcome correspondence from those they have been unable to trace.

GOVERNMENT BY NATURAL SELECTION

BY

HUGH TAYLOR

METHUEN & CO. LTD.
36 ESSEX STREET W.C.
LONDON

First Published in 1915

PREFACE

THE course of politics during the last twelve months has once more emphasized a truth which is constantly reaffirmed in the pages of history. When a nation finds itself confronted by some great and imminent danger, it is compelled, if it is wise, to cease trifling with political whims and constitutional fancies, and to restrict itself to the requirements essential for national salvation. *In periculo veritas:* the cry of distress is the cry of truth, and the demand for an administration of the best available men which is ringing through the country to-day represents the most fundamental principle of national health and the indispensable condition of sound political evolution. The fact that such a demand has been made is in complete accordance with the requirements of the present theory. Some of the judgments and anticipations recorded in these pages might indeed seem to be a mere generalization from contemporary conditions. The author therefore owes it to himself and his readers to state that the work was entirely finished in its present form, except for some literary alterations, more than a year before the declaration of war, and was in most respects substantially complete many years earlier.

CONTENTS

CHAPTER VI

GOVERNMENT BY NATURAL SELECTION

GOVERNMENT BY NATURAL SELECTION

CHAPTER I

INTRODUCTION

IT is not without a wholesome sense of his own temerity that the author ventures to suggest a new point of view from which to study the phenomena of history. Yet perhaps some considerations may be urged in extenuation of an enterprise which, to those who are aware of the difficulties involved, may seem hazardous in the extreme. Thirty years have now elapsed since Sir John Seeley endeavoured to enforce the truth that history is something more than vivid or picturesque narrative. The rising generation to which he spoke is now the departing generation; yet both his appeal and his example seem to have been without effect, since nothing of note has been done towards the effective inauguration of a real historical science. Indeed with many, though not with all historians, absolute scepticism still prevails as to the very possibility of discovering a principle which shall serve to disentangle the issues or unravel the complexity of history. In spite of the fact that science has extended its triumphal progress in so many directions, the realm of history is still unconquered. From the strictly scientific point of view we neither know what is taking place nor how it is taking place. We do not know what is taking place, because a mere description of historical events, however bright and animated, gives us no real insight into the actual nature of the panorama which is thus made to pass before our eyes. We do not know how it is taking place because we are ignorant of the real causes of some of the most significant developments in human affairs. Only vague suggestions, for instance, are made in explanation of national success and national failure, perhaps the most important phenomena in the whole history of the human race. In despair, apparently, of solving problems so intricate, historians busy themselves with the endless multiplication or restatement of facts; and leading thinkers are still found to assert that, if we add the right to inculcate certain moral lessons, history never has had, and never will have, any other legitimate occupation. There arises in consequence a very general disposition to acquiesce in the conclusion that history contains no secret which has not already been revealed; and until this orthodox and conventional attitude is abandoned in favour of more enterprising methods, until we witness a change of the spirit in which the whole subject is approached, further progress seems absolutely impossible.

It is, of course, natural for laborious and earnest-minded inquirers to believe that, if they themselves have spent their lives in historical research, and have discovered nothing new, therefore there is nothing new to be discovered. Human knowledge would have, however, remained in its first infancy had such an attitude been adopted with regard to learning in general by those theoretically engaged in its advancement. Yet there can be no doubt that the professional historians, who constitute what is called authority, still frown on any attempted originality in the study of history: and authority so constituted predetermines and prescribes the lines on which alone investigation should be pursued. This is an assertion of which abundant proof will be given in the succeeding pages, when the want of real evidence for many authoritative opinions will be shown to coexist with the strongest possible disinclination to relinquish them. This conservatism is no doubt intensified by an instinctive perception of the fact that alteration in our views of history means alteration in our views of life. In this respect, indeed, there is a vast and unfortunate difference between history and the other sciences. A new theory in chemistry leaves individual idiosyncrasies unaffected; a new theory in history touches the most vital prejudices. The greatest repugnance, for instance, is manifested to the readjustment of political values, because to alter the theory of the State is to affect the life of the individual. Yet it is right that established constitutional views should be questioned, if only to discover whether a mistaken estimate of the relative value of political forces is not one of the causes of the unprogressive nature of historical science.

But if the accredited leaders of historical investigation do nothing to remove the reproach which rests upon a barren and unproductive study, there is the more excuse if those who can in no way claim to belong to the ranks of professional historians should feel compelled to make the attempt. For such an effort on the part of somebody there would seem to be a very pressing need. If history were some form of purely intellectual pursuit of more than ordinary unattractiveness, and which promised no material reward, it might have been possible to comprehend the prevailing apathy with regard to the rich possibilities which may result from a successful originality of method. But the tame acquiescence of the majority of historians in the conclusion that the last word has been said upon the conduct of historical research is without even this excuse. To those who regard the advancement of humanity as a sufficient object of their endeavours, history offers the greatest of all possible rewards. None of the previous triumphs of science can compare in beneficence with those which will ensue from an intelligible explanation of the course of political evolution. Success in the elucidation of certain historical problems will reform not merely the theory of history, but eventually the practice of politics as well. At present the inconclusive nature of the lessons drawn from history is reflected in the bewilderment of contemporary political life, with its chaos of contending principles and irreconcilable theories. All this might be replaced by some steady endeavour towards some definite end, if the riddle of past ages could be read aright: but until this is done with some approach to success, all efforts at reform will continue to be characterized by their present incoherence. No attempts at social or political regeneration can meet with permanent success unless they conform to the laws or tendencies which are implicit in the movements of history. The discovery of these laws and tendencies would seem, therefore, to us a matter of surpassing importance. In spite of all discouragement, the younger generation in particular should never allow themselves to forget that the pages of past history rightly interpreted contain the secret of all future political progress.

The study of history, it is true, has not as yet revealed any principle which is not the subject of endless dispute. With regard to this absence of fruitful results, the well-known words of the great founder of modern scientific method are still worthy of consideration. "If our study of science be thus barren," said Bacon, "our method of study must be wrong." Exactly the same argument which applied to the treatment of science in general at the time of Bacon will apply to the treatment of history at the present day. If the results of the present study of history are, as the unprejudiced observer must undoubtedly conclude, "thus barren," our method of study must be wrong. There are those, however, who differ from Bacon as to what the end and aim of research should be, and who think that the admission of a practical object in historical investigation must involve a fatal departure from the principles of true scientific procedure. But in the first place it may be remarked that though Bacon apparently forfeited the respect of these philosophers by pursuing a practical object, yet the actual result of his teaching was an increase of the true spirit of science. In the second place, with reference to the study of history, those who regard the admission of a practical object as subversive of true scientific method mistake the problem presented by political evolution. The statesman, like the physician, must know the nature of the organism for which he prescribes. Yet such is our want of real knowledge that the theory and practice of politics at the present day is in much the same condition as was the theory and practice of medicine before the discovery of the circulation of the blood. It might perhaps be too much to say that in the art of politics, as in the art of medicine, man can do no more than assist nature. But this at least is certain, that the art of government cannot produce the best results until our knowledge is such that all schemes of political improvement may be subordinated to the fundamental principles of political evolution. Accordingly, since all that is necessary for this purpose is an acquaintance with the actual process of nature, it follows that the hope of practical results may be entertained consistently with the retention of the utmost severity of scientific method. The really scientific historian must necessarily assist the art of practical politics whether he wishes to do so or not: for it is upon the ascertainment of the original and permanent political tendencies of man that a true art and science of politics can alone be based.

We may conclude, then, with Bacon that meagre results prove the existence of a faulty method; and after a study of the procedure which has brought success to other sciences, it is not difficult to suggest the lines upon which alone improvement can be effected. In the first place historians seem to have fallen into the habit of accumulating facts as a miser accumulates gold, not for any ulterior purpose but merely to satisfy the desire of acquisition. The increase of the hoard is an end in itself; and those who so increase it are honoured for the mere extent of their learning, apparently without consideration for the uses to which that learning should be put. It is not, as yet, sufficiently understood that a life-long and laborious acquisition of facts may be a positive disqualification for the production of really scientific results. In the case of men like the late Lord Acton an extensive acquaintance with the most minute circumstances of a given period, and a too complete knowledge of the many-sidedness of any given event, seems to produce a sort of paralysis of the constructive faculties, which prevents them from making any real and effective use of the vast array of materials collected in the storehouse of their mind. To put the objection in its simplest form, such writers do not know what to omit, or what to emphasize, but leave the reader helpless and without guidance amid masses of conscientiously accumulated information; and what is worse, they and their admirers deny the right of any one to attempt the interpretation of history who is not equipped or, rather, overweighted with a similarly encyclopædic knowledge. The danger at the present day arises not from pretentious ignorance, but from unenterprising erudition. It would seem abundantly possible that the mind may become so choked with detail as to be incapable of the distinctive intellectual operations which constitute a really scientific method of procedure. It is clear, therefore, that in this ceaseless effort to increase the already overwhelming mass of historical facts, without the suggestion of some new and vitalizing principle, the method of history is wrong. The reform of all other sciences has taken place in consequence of the introduction of some leading idea, around which phenomena have been grouped in orderly array, and the reform of history must be accomplished in the same way.

It has indeed already been recognized by many thinkers that this is the direction in which we must look for any real progress; yet failure has resulted from the adoption of this method also, and for the following reason.—Those reformers of historical method who have taken as their leading idea such notions as justice, liberty, self-government, or "the immortal working of the moral powers," have made the mistake of choosing their unifying principle from among conditions of civilization which are too advanced. What is really wanted is some motive more in keeping with the lowly origin of man, an impulse which can be shown in operation at the earliest as well as at the latest stages of history, and which plays its part in political life from the beginning to the end of social evolution. The higher aspirations of the human mind, though they are in a measure realized in the course of human progress, do not on that account afford any necessary indication of the method by which such progress has been achieved. Philosophers have erred by thinking too highly of the political nature of man, and have in consequence been inclined to over-estimate the complexity of the motive to which he has owed his political advance. Simplicity is the secret of the successful explanation of even the most intricate phenomena, at least at the beginning of an investigation. The prevailing instincts of man are simple even at the present day, and must have been still more simple in the past, and it is in the simple rather than in the complex impulses that the cause of political progress must be sought.

In the third place it is generally assumed that political progress has been due entirely to the conscious efforts of man to secure for himself a greater share of social happiness, and a more reasonable scheme of political existence. With this end in view, he has framed an endless variety of constitutions with more or less success, and to these deliberate attempts to modify governmental conditions in his own favour the political progress of the race is thought to be entirely due. In the opinion of the present author, however, the transition of man from a savage and even bestial state to reasonable political surroundings is only intelligible on the assumption that he has received assistance from some unconscious tendency of nature. The solution of the riddle of history will accordingly depend, not on the minute chronicling and description of the various efforts of man to modify political conditions, but in a discovery of the chief of those unconscious tendencies which are directing society to an unknown end. Assuming, then, the existence of some such tendency, the direction in which we may look for its discovery with most hope of success seems clearly indicated by the previous course of science. It is in the highest degree antecedently probable that there is some connexion between Darwinian principles and the development of man in societies. Some modification of the Darwinian struggle for existence will in all probability give us the clue of which we are in search. It would indeed be strange that, when all the other sciences connected with the development of life have been radically altered by the Darwinian theory, its bearing on history should not be equally intimate and important. If, then, the co-operation of some Darwinian principle be admitted, this implies the abandonment, as already pointed out, of the high philosophic standpoint. Instead of seeking in history the realization of metaphysical ideas, the exemplification of some Hegelian principle, or the discovery of a mysterious something called a "psychical diapason," we must be content with an evolutional impulse more befitting the humble origin of an animal who in the time of Aristotle may have deserved the term political, but whose political adaptability is itself the result of tendencies that were in operation long before the very rudiments of social life were possible.

But, having decided on the application of Darwinian principles to history, the practical question arises, at what point of a somewhat serious and extensive undertaking to begin. Naturally the reader will say at the beginning. Yet though investigation in such matters must begin at the beginning, it does not follow that the exposition should do likewise. That which comes first in the process of inquiry does not necessarily come first in the explanation of the results. In the present instance the desire of rendering the argument clear and intelligible, as well as the necessity of restricting the length of the work, has rendered impracticable the full and comprehensive treatment of the subject from beginning to end. In consequence of these and other considerations it has seemed advisable to the author to commence what he has to say in the middle, and to give the later portion of the theory first. The present work, however, though only a part of the original scheme, is designed nevertheless to form a connected whole, which may be accepted or rejected on its merits. The discussion will deal only with the evolution of government subsequent to the institution of monarchy or of some fixed political control, which is assumed as existing before the present inquiry begins. That portion of the theory which treats of early political evolution before government has become firmly established will be reserved for a subsequent volume, entitled "The Origin of Government." It will there be shown that the evidence for the existence of a law tending in the direction of government by natural selection is much more extensive, and covers a much wider range of existence, than can be indicated in a treatise which is restricted to the later manifestations of a principle which throughout all ages has inspired the most efficient rulers and statesmen.

Not long ago a learned professor and a smart historian, while reviewing the course of historical study in the last half century, both took occasion to exult in the apparent discomfiture of those who have ventured to believe in the existence of some great law or tendency which should serve to unify the study of history. To such reasoners, of a type that would have aroused the intellectual scorn of Bacon, it is impossible to reply except by asserting that however long an interval of fruitless effort may elapse, whether it be ten or fifty or a hundred years, some such principle will one day be unquestionably discovered. For the failure of the search up to the present day, cogent reasons have already been assigned. But to these may now be added a further consideration of greater importance than any previously urged. No real progress will be made in historical science until we learn to look at history with new eyes, and approach the task with a determination to shake off the numbing influence of mere habit and custom, upon the intelligence of the observer. In the mind of the inquirer there must, in fact, be a readiness to allow the possibility of a new interpretation for some of the oldest and most commonplace events. Paradoxical as it may sound, it is in truth a too intimate acquaintance with the everyday aspect of certain historical happenings which prevents the perception of the deeper meaning which in reality they possess. Familiarity breeds contempt, and leads us to underrate unduly the import of circumstances about which we have read too much and which we know too well. Or to state the case with more precision, unobservant familiarity with certain incessantly repeated phenomena breeds indifference to their possibly far-reaching significance. Even that most encouraging and helpful of thinkers, Professor Bury, seems to undervalue the importance of our actually accessible data. Though confident that at any moment a revelation may be made which will revolutionize the whole study of history, Professor Bury's courage does not extend to the belief that an idea of such immense presumptive importance can be discovered among the already available materials of history. Under the conviction that the facts with which we are familiar have already yielded all their meaning, he urges inquirers to leave no historical hole or cranny unexplored, in the hope that the great secret may be accidentally found lurking in some abstruse account of some unfamiliar people. But if we may judge by the way in which other sciences have been reformed, the great historical truth of which we are in search will not be found in some obscure corner, nor will the great discovery be made by ransacking unlikely records. On the contrary, nothing is more certain than that the idea which is to transform the study of history, whatever that idea may be, is lying somewhere unrecognized on the pages of those very text-books with which we have been familiar from childhood. To find a new significance in old and well-known facts is the very essence of the scientific process.

CHAPTER II

GOVERNMENT BY NATURAL SELECTION

PROFESSOR HUXLEY, in discussing the relation of the Darwinian theory to human life, and the impossibility of applying that theory directly to social conditions, gives it as his opinion that the struggle for existence among animals has been replaced by the struggle for subsistence among human beings. When Huxley wrote this he was giving expression to an opinion which is still very generally prevalent. But he entirely missed the fact that together with the struggle for subsistence there is a second and perhaps more important struggle going on simultaneously. In every civilized State there are many active and ambitious men whose position is assured from birth: there are many others whose capacity for rapidly acquiring wealth is so remarkable that they have practically never been within the reach of want, and who cannot therefore be regarded as engaged in a struggle to avoid it. Among such men there is a sufficiently large number of individuals marked by a certain superiority of character: their requirements are intellectual rather than material: and the rivalry upon which they instinctively enter has always had for its object something far higher and more important than the mere struggle for subsistence, or even than the struggle for the increased means of enjoyment which the possession of wealth gives.

Under generalizations about the competition of life two distinct processes are almost habitually confused. In every progressive community those individuals whose character and intelligence is above the average are engaged, not in a struggle for subsistence, but in a competition for reputation and honour. With all the most active and able members of society, success in the struggle for subsistence or even in the pursuit of wealth is not the end and aim of their most earnest efforts, but is merely preliminary to the gratification of ulterior objects of ambition. The fact that competition for the distinctions of life is a motive which under certain circumstances supersedes the struggle for subsistence can be proved not only in the case of many born to high position, but also from the career of those successful men who have raised themselves by their own exertions. The prosperous merchant despises the modest competence which formed the humble ideal of his earliest aspirations, and is not even content with the acquisition of that wealth which was the goal of his later endeavours. If he has been sufficiently clever or sufficiently fortunate to achieve pecuniary success before the prime of life is passed, his next step is almost invariably to secure some administrative post in his native town, or to enlarge the sphere of his ambition by entering Parliament. The struggle for subsistence or for wealth is not what Huxley thought it, the main feature of the game of life, but only the preliminary round which entitles the survivor to enter the final stages of a more important competition. Men of real character in a progressive nation have never regarded wealth as a means of self-indulgence, but merely as a step towards the achievement of some more honourable title to consideration; and history shows us that in all ages great intellectual capacity has found its deepest satisfaction in the pursuit and exercise of political power. There is, in fact, ample evidence in history to suggest that at a certain stage of civilization the struggle for existence is replaced by another great law of nature, in accordance with which under competitive conditions political distinction is most eagerly sought by those individuals who excel in mental capacity and strength of character. It is the object of the present work to show that this rivalry for political distinction and political power is an evolutional impulse of the same character as the struggle for existence itself, and that it has in addition the effect of inducing the ablest members of a nation to place their talent or their genius exactly where it is of the most surpassing importance, namely, in the governmental service of the State.

Though aware that the study of history does not

encourage any tendency to idealism in politics, the moralist will still perhaps be inclined to resent the systematic construction of a theory based upon the assertion that the desire for distinction has been the predominating motive in the conduct of statesmen. Government is perhaps the most serious of all human undertakings. It is natural, therefore, to assume that this transcendent duty must necessarily have appeared as solemn and as serious to the statesmen who have come forward to perform the task, as to those philosophers and moralists who indicate the spirit in which it ought to be performed. History does not confirm this view. During periods of political disturbance, when the welfare of the community appears to be seriously threatened, cases of unselfish devotion do certainly occur, and these will be presently explained. But except in times of revolution, or during periods when the normal life of the community is disturbed, the inspiring motive of a great political career must be sought elsewhere than in a sense of duty. Gibbon was of the opinion that among monarchs the reign of the Antonines was "the only period of history in which the happiness of a great people was the sole object of government." The record of statesmen may be somewhat better, but at the same time a study of biography reveals the fact that their actions have certainly not been inspired by unalloyed desire for the public welfare. In an inductive inquiry like the present, which is directed not to the improvement of morals but to the ascertainment of facts, the subject assumes a purely historic and scientific aspect, and the question, therefore, to be decided is not the nobility but the efficacy of the motive which has induced men of talent to devote themselves to public life. A theory of politics which asserts the prevalence and the value of motives of ambition cannot be disproved by demonstrating its moral deficiencies, but must be judged at the bar of history alone. Finally, it may be observed that only in the philosopher's study is the desire for distinction regarded as something irreconcilable with the desire of doing good. The one impulse may have a less selfish origin than the other, but this obvious difference is magnified by moralists into a fanciful and exaggerated opposition, unknown to history and untrue to life. The supposed antipathy between them is no record of general experience; it is the outcome not of the scientific impulse to describe human nature, but of the moral longing to improve it. To covet honour has never been regarded as a sin or even as a weakness in real life. The desire for a great career does not exclude the simultaneous operation of higher motives, and in all but saintly characters, with which we are not concerned, ambition, noble perhaps but still personal, has supplied the inspiration of the most famous imperial and political achievements. To read their history in a nation's eyes has at all times rightly been regarded by practical men as the legitimate aim and the fitting reward of even the greatest statesmen.

But while asserting that this is an accurate account of the inducements which have led the more distinguished statesmen to engage in politics, it is unnecessary to deny that instances of unselfish political conduct are not infrequently to be found upon the pages of history. Such an admission is in no way inconsistent with the position just laid down. The present inquiry is mainly concerned with the motives which inspire the action of those who govern. The unselfish conduct of the patriotic reformer is inspired by motives which affect those who are governed, a truth which the argument of the succeeding pages will tend to make more clear. The impulse by which the Gracchi are urged forward on behalf of their distressed fellow-countrymen, or the Pyms and the Hampdens against the unjust pretensions of monarchy, must in the last analysis be classed as a self-protective instinct of the individual citizen, which leads him to oppose a course of governmental action which he believes to be highly dangerous to himself and the community. As champions of the nation against unjust rulers, reforming patriots play a distinctive part in the evolution of government, which will be better apprehended when the true meaning of representation has been explained. Their motives, however, are not those with which politicians are normally inspired: strictly speaking they are engaged not in conducting government, but in reforming its conduct. It is with the everyday work of an administration, engaged in keeping the people contented at home and in furthering the interests of the nation abroad, that the present argument deals, and this is performed by statesmen not in a spirit of self-sacrifice, but for the motives already described, which are not the less patriotically efficient because they may happen to be less exalted.

If, however, we conclude that the desire for power or pre-eminence has been the ruling motive of statesmen, we are not on that account bound to deplore the fact as a symptom of the weakness and wickedness of human nature. On the contrary, it is possible to hold that this universal spirit of rivalry is the inspiring cause of an evolutional process of the deepest and most far-reaching significance, whereby those who are exceptionally gifted with intellectual power, moral determination, and a knowledge of human nature, all of them qualities useful in government, tend under favourable circumstances to find their way to positions of the greatest political influence. According to the present theory this natural desire for patriotic distinction is a feature of political evolution which intimately affects the vitality of the State, and with which it is dangerous and even fatal to interfere. One of the most essential conditions of healthy political life is that the ablest members of the community should be allowed to gratify their love of distinction in the service of the State.

Compliance with this inherent tendency of nature is of the greatest political advantage, because it secures a continuous supply of intellect to reinforce that most vital of all the national functions, the function of government. It will be argued in the following pages that the success or failure of nations has been proportionate to the freedom with which this principle has been allowed to influence the character of the national administration, and that the utilization by the social organism of its highest individual talent for its own political advantage is a condition of continued national success. The triumph of the most successful nations in history is inseparably connected with the facilities offered for the political predominance of great or talented men. Realization of this principle constitutes a kind of government by natural selection, a process by which those who are approximately the fittest find their way by their own natural ability to the most important governmental posts. The State is thus enabled in times both of internal difficulty and external danger to enjoy the guidance of its ablest citizens, whose personality tends to influence for good the political institutions and even the character of the people, while the mere memory of their names in after ages is a perpetual national inspiration. The keenest intellects among all those to whom the love of power appeals are brought to bear upon the perplexing problems of social existence, and the success of the community is assured in proportion as the decisions of sagacious statesmanship are systematically allowed to influence the life and policy of the nation. And because this is a natural process, its interruption is followed by a natural and inevitable penalty. If as a consequence of royal incapacity, aristocratic privilege, or democratic ignorance, the State is deprived of the guidance of that succession of able rulers which under more favourable circumstances would have been naturally available, it falls into weakness and decay, begins to exhibit all the symptoms of national decline, and is finally left at the mercy of strong and aggressive rivals. Though it is true that the full advantages of administrative genius cannot be secured without the co-operation of certain other political factors, of which the absence may even be sufficient to endanger the existence of the State, yet it may unhesitatingly be affirmed that government by ability is the most indispensable of all the conditions of national well-being; and that no nation can permanently hope to succeed unless its constitutional arrangements are such as to favour the operation of the process which has here been described as government by natural selection.

If we turn to the history of Rome with this idea in our minds, that whatever different constitutional forms a State may assume, the most essential requisite is the encouragement of ability to find its way to office, it is remarkable what a light is thereby thrown upon the elusive secret of Roman greatness. Amid all the injustice inseparable from an imperfect conception of political duties, such as was characteristic of this early stage of social evolution, there was this conspicuous merit in the Roman system, that no constitutional disabilities prevented the attainment of the highest prizes of a political career by any deserving individual among that remarkable oligarchy. The highly cultivated members of an almost endless succession of noble families incessantly competed with one another for the graduated offices of the State, thus keeping up within the ranks of government that constant supply of ability which made Rome mistress of the world. In a way never afterwards even approximately realized until we come to the history of England, the talent of the leading men of the nation was enlisted in the service of the government, undergoing a competitive training as they filled, one after another, the various offices of the State. The system was not ideally perfect, even as a method of eliminating incapacity. Yet the great men who were allowed in consequence of sheer merit to attain the highest political positions, have left a record of achievement unequalled in the history of the world. They rose to these positions by free competition for political honour and distinction, in obedience to a natural impulse which in other communities has, on various pretexts, been hampered and restricted, but to which the Romans so wisely gave full play. With that unconscious genius which marks so much of their political procedure they managed to make even the prestige of aristocratic distinction incomplete in the absence of achievements which augmented the greatness of the Republic. The pride of birth and the pride of wealth, faults which have ruined other nations, were subordinated by this extraordinary race to the nobler pride of work performed and victories won in the service of the State. With all the grievous faults and sins of Roman administration, the enlistment of the utmost energies of its ablest individuals for the purpose of building up Imperial greatness was never more thoroughly accomplished in the whole previous or subsequent history of mankind.

The success of those States in which the political system is so arranged as to ensure the supremacy of the ablest and most intellectually distinguished citizens is similarly illustrated by the career of Athens. We have all been taught to believe that it was the element of freedom in her institutions that made Athens great: we are, in fact, told that "freedom" makes all States great, but of the way in which it operates to produce this result we are given only the vaguest and most unsatisfactory account. The author hopes to show before the conclusion of this work that "free institutions," whatever may be their moral value to a certain number of self-restrained individuals, are of advantage to the State and to the general community only so long as they facilitate and encourage the predominance of political genius, but that otherwise their inauguration partakes of the nature of a dangerous and even fatal experiment, as was very speedily apparent in the case of Athens herself. It is indeed not a little remarkable that in spite of the permanent advantages that political freedom is supposed to impart, the brilliant period of the political greatness of the Athenian Republic coincides only with the supremacy of her greatest men, and departs with their departure. So long as the temper of the people facilitated the triumph of individual greatness, the State displayed qualities of surpassing brilliance; when the political conditions were gradually altered, so that power fell into the hands of inferior men, her greatness was abruptly terminated, and neither the principle of liberty nor the habit of free discussion, nor any of the other advantages which are supposed to belong to a free constitution, were able to prevent the headlong ruin of the State.

Amid the vicissitudes which the Athenian constitution underwent from the time of Solon to that of Pericles, that one feature which is here regarded as fundamental, while all the rest are accessory, remained practically unchanged. The access of the ablest men to office was, during this period virtually, if not actually and deliberately, fostered. Under the constitution of Solon the Eupatridæ or nobles were encouraged, says Oman, " to administer the State under due control and for the benefit of the whole community,"[1] an aim upon which the greatest refinement of political science has not yet

[1] Oman's "Greece," p. 109.

been able to improve. Again, the first effect of the reforms of Cleisthenes was undoubtedly to increase the chances of the rise of great men to power, by encouraging the principle of natural selection in politics, and by widening the area over which that selection took place. The key-note of his system was the abolition of privilege, and to abolish privilege is, from the point of view here adopted, to assign political power in proportion to political merit. His measures were directed to the object of interesting the whole body of citizens in the welfare of the State, and the open career for political talent became the order of the day. The ablest men were thus encouraged to reach, and as a matter of fact did reach, those positions of administrative power for which nature had intended them. By the institution of the famous custom of ostracism, so much discussed and so little understood, the Athenians practically recognized the struggle for supremacy as a principle of politics; and while taking care to retain all the advantages of the process, they hit upon the method of ostracism to minimize the evil, by bringing the struggle to a speedy termination, and leaving one of the rivals in control of the policy of the State. To these brief remarks upon the working of government by natural selection in the Athenian State more will be added in the succeeding chapter, and again in the chapter on "Democracy." Additional arguments will there be adduced to show that the great secret of the political value of liberty lies in the fact that it promotes the access of the ablest men to power. When it ceases to produce this effect its health-giving qualities disappear, and it tends to produce political conditions which are inconsistent with the efficacy of government and the reasonable subordination of individual aims to necessary national ends. It is not, of course, contended that all nations can be made great by the simple process of putting their ablest citizens in office, or that the political predominance of statesmen like Themistocles or Pericles is alone sufficient to account for Athenian greatness. There is undoubtedly a different potential value in the qualities of different races, and even of different States, which predestines some to a higher place in the annals of history than can be attained by others. But though political selection may in two different countries have very different material upon which to work, government by ability produces the best results attainable under the given circumstances. Granting that the Romans and the Athenians seem to have been endowed from the first with qualities for which no government theory will account, it is equally certain that their appreciation and utilization of individual talent in politics raised them to a degree of greatness which they would not otherwise have attained.

In the history of the French after the time of Richelieu we have an illustration of the opposite truth, that when those best qualified by hereditary characteristics and natural genius to guide the policy of the State, are nevertheless prevented from having access to the ranks of government, misdirection of the national energies together with all the inevitable evils or perverse and unintelligent administration begin to impair the vitality of the nation and to ensure its inevitable decline. It is the essence of the present theory that no State can hope for continued prosperity unless its constitution is such as to facilitate the rise of ability to political power. The history of France is the history of incessant failure to realize this essential condition of steady political progress. Whatever other causes may have contributed to the misfortunes which culminated in the outbreak of the French revolution, the radical disorder from which the nation was suffering is unquestionably to be found in the royal monopoly of power, which caused the unjust and injudicious exclusion of a talented nobility from that sphere which forms a proper and natural outlet for their ambition, namely, the service of the State.

The popular belief with regard to this great political cataclysm is that it was caused by the insolent demeanour and unjust privileges of the aristocracy, who had constituted themselves a separate and almost sacred caste, distinct from the rest of the nation. The more scientific view may be expressed in the words of Bright, " that the explosion had been slowly preparing since Louis XIV. had completed the mistaken policy of centralization and had been able to say that the King and the State were one." Both these views contain an element of the truth, but neither trace the evil to its real source. The unjust privileges and exclusive demeanour of the nobility were the result of other grave faults in the political system, which will be presently examined, and they cannot with justice be regarded as the principal cause of the national disorders. Again, with regard to the second alleged cause of the revolution, that it was due to a mistaken policy of centralization, this theory, though very generally accepted, does not supply a really full and adequate explanation of the disasters which fell upon France. No doubt excessive centralization is always to a certain extent an evil, but surely in the present case the decisive feature of the situation is to be found not in centralization alone, but in the additional circumstance that the government was centralized in the painfully inefficient hands of Louis XIV. and of Louis XV. The ruin of France under Louis XIV. was not caused by the fact that the government was in the hands of one man, but by the fact that it was in the hands of one man who was vainglorious, jealous, and as a statesman immeasurably stupid. If Louis XIV. had been a Frederick the Great, it would have mattered very little that the government was centralized. The pitiful penury which resulted from the criminal mismanagement of the resources of the State in unsuccessful warfare would then have been avoided, and France would have been made sufficiently great and prosperous, even without a constitutional government, to escape the fiscal iniquities and the financial makeshifts which were by far the most potent of the secondary causes of the revolution. In that case the revolution itself would not have taken place or not in so disastrous a form, and those additional guarantees of good government, which must necessarily supplement wise statesmanship, would gradually have been supplied by a process of comparatively peaceful reform. The really vital point with which we are here dealing is not whether political power was unduly concentrated, but what was the character of the authority in whose hands that power was placed. Was the intellectual quality of the guiding mind of the nation equal to its supremely important task ? It is the competence of an administration, centralized or not, that matters, and one of the very best ways of failing to secure competent administration is to have an hereditary monarchy where the king not only rules but governs. The source of all the ruinous misfortunes of the reign of Louis XIV. was not primarily the incessant war which impoverished the country, nor, as will be shown in the succeeding chapter, want of a representative system, nor the iniquitous system of taxation, nor any of the causes usually assigned, which are in reality only secondary. It was the claim, the fatal hereditary claim, of an earnest but stupid king to perform political duties for which he was intellectually unfitted. It is where the misdirected but all-pervading activity of an inferior monarch excludes his ablest subjects from the administration, that the worst evils of a centralized autocracy are experienced. Wherever Louis XIV. meddled in his royal and pathetically ignorant conceit, and he meddled everywhere, disaster followed. We hear much of the injury which is inflicted on the moral character of individuals by the denial of liberty. We hear little or nothing of the injury inflicted on the intellectual character of government under similar circumstances. The greatest evil from which France suffered at this time was not the suppression of freedom, but the supersession of genius: not the denial of the rights of the individual, so much as the denial of the right of talent to contribute its proper share to the moulding of the destinies of the nation: not the scandalous drain on the national income, but a defect in the constitution which was more costly still. The greatest potential wealth of a nation lies in its hidden springs of administrative ability. If the constitution is so framed as to hinder the emergence of political talent, the resulting waste of the intellectual resources of the State inflicts a more serious injury on the community than the most reckless dissipation of material wealth. "The day of third-rate statesmen was come: and Louis, self-satisfied, ignorant, and unable to distinguish between a good and bad minister—that surest mark of an incompetent ruler—flattered himself that he could form statesmen as he would, and preferred, with fatal security, a docile mediocrity to the dangers of originality and power."[1]

To the loss incurred by the denial of a career to political talent, and the consequent impoverishment of the intellectual vitality of the administration, must be added the mischief which arises when an ambitious and energetic class find themselves deprived of that outlet for their energies which is at once most natural and most useful. The nobles, instead of being employed in honourable political duties, were encouraged, even compelled, to waste their time and substance in riotous living in the capital; and those distinctions which should be the reward of honourable service were distributed at the mere caprice of the monarch, since there was no honourable service to reward. If the constitutional arrangements and the accompanying political distinctions which it is in the power of the State to bestow are so adjusted as to encourage the most active and energetic of the nobility to give the best of their talent and intellect to the service of the community, the secret of good government, at a certain stage of political evolution, has to a large extent been solved. The success of Rome in this respect is in marked contrast to the failure of France. In the aristocratic classes at a certain period of constitutional development are to be found the strength of character, the knowledge of men, and, it may be added, the personal pride and ambition which constitute a unique claim to the

[1] Kitchin's "France," vol. iii. p. 239.

exercise of the functions of government. This class and their characteristics are valuable, even indispensable, political material, and to suppress them or even to neglect them is a ruinous policy for any nation to pursue. The exclusion of the French nobility from the political sphere gave a useless and pernicious direction to their personal rivalry and private ambitions, and thus contributed to the ruin of the country and nobility alike. The neglect of their obvious duties by the landed aristocracy, which unquestionably both hastened and embittered the revolution, was itself inevitable in view of the fatal policy of Louis XIV., who deliberately discouraged any interest in politics or in the social and economic welfare of the nation, on the ground that it interfered with a sphere of influence reserved for himself alone.

While endeavouring to save the State from the dangers which are threatened by the existence of a too powerful nobility, the rulers of France before this period had incurred other dangers of a less obvious but of a no less fatal character. When they should merely have restrained the turbulence of the nobility, they achieved their political annihilation. But the vigorous development of the power of a State depends on the due subordination, not on the annihilation, of the ambitions of an aristocracy. The French, like the Romans, should have been encouraged to believe that political distinction was a legitimate ground for pride, and that mutual rivalries are best settled in the service of the State. How different was the policy actually pursued by Louis is only too well known. "Louis was unwearied in his watchfulness lest any one should become powerful in France, or have even the shadow of independence. The *noblesse*, grouped round his person, were taught to sell themselves for the pomps and fêtes of court; in these they wore away their wealth and grew yearly less able to assert themselves."[1] We turn contemptuously from the numerous and well-authenticated accounts of the struggle for social precedence and privilege which marked the relations of the French nobility to one another. Yet these unhappy bickerings were the natural and inevitable result of the policy that has been described. The ambitions and rivalries of individuals are not the merely deplorable features of a weak and erring humanity, which moralists in general would have us believe, but an essential part of the process of political evolution. They are the mark of a progressive nation, and work mischief only if denied a proper outlet. The French nobility, forbidden to set foot upon the political arena, where their rivalry might, in Roman fashion, have contributed to the greatness of the State, turned instead to a pitiable struggle for social precedence, and an unworthy competition for the empty favours of the monarch. The consequence was, that though occasionally real genius, like that of Colbert and Louvois, was allowed to take part in the business of

[1] Kitchin's "France," vol. iii. p. 163.

the State, yet the certain and regular supply of talent, which a system of free political selection is alone able to provide, was never at the disposal of Government; while even the genius of such men as Colbert ar 1 Louvois was largely neutralized by the caprice or ignorance of the master whom they served. The occasional brilliance of individual ability produced no enduring result for France, for the simple reason that the isolated efforts of the most gifted subordinates are insufficient to counteract the mischief which inevitably results from the hereditary right of an ignorant monarch to misdirect every detail of administration.

The harsh and injudicious suppression of the political activity of the nobles in France forms a remarkable contrast with the system which grew up in England, where, from comparatively early times, the more prominent members of the English nobility were in the habit of taking part in the government of the country. A constant supply of ability thus reached the administration, and to this cause the striking superiority of the English government, and the happier national fortunes of the English people, are in a very large degree to be attributed. From the time, practically speaking, of William the Conqueror, the political subordination of the nobility was secured, while leaving their vitality unimpaired. But a further step was needed before the full potential usefulness of the English aristocracy could be realized in political action. The old pretension of the monarchy to rule in accordance with its own views, and in virtue of its ancient title to political supremacy, still stood to a certain extent in the way of decisive political advance. The hereditary king, who not only rules but governs, is an insuperable barrier in the way of the accession of the finest talent of the nation to administrative power. While he remains, the intellectual quality of the government is at the mercy of the accident of birth, which may at any time invest dull and perverse mediocrity with an inalienable right to misdirect the nation's course. A profoundly important but almost wholly unrecognized feature of the change from absolute to constitutional monarchy consists in the transference of the power which directs the activities of the State from an hereditary king who may be incompetent, to a statesman who, if he has been permitted to raise himself by his own exertions, is sure to be the reverse. Nor is this truth in the least invalidated by the fact that kings have occasionally in history happened also to be great statesmen. Such an eventuality is of necessity the merest accident, and the quality of the government of a State, if it is to survive in the international struggle for existence, cannot be left to accident. When, however, monarchy is duly restricted, then the natural desire for political power or distinction, released from the bonds which formerly held it in repression, irresistibly urges talent and genius to the head of the national administration. Throughout English history, in proportion as the control of the national destinies has been withdrawn from the hereditary monarch, it has tended to fall into the hands of one of the ablest of his subjects. Under constitutional monarchy the permission which is accorded to the wealthy and educated classes to engage in public life has the effect of bringing forward the most active and ambitious of their members in quest of that political influence which has been so strangely valued throughout the history of the human race, and the chief condition of political efficiency and national progress, as laid down by the present theory, is as a natural consequence realized. Whenever as the result of constitutional restrictions a great justiciar or minister has practically assumed the administration of the realm, the process of government by natural selection has virtually been in operation. We can, however, assign a date when these evolutional tendencies seem to find a definite fulfilment; and though the recognition of the value of government by ability does not even then become quite conscious and deliberate, nevertheless it was the revolution of 1688 which rendered systematic the supply of that individual efficiency upon which the very existence of a nation may from time to time depend.

The chief importance of the revolution of 1688 is generally held to consist in the enforcement of the conclusion that the king rules by the will of the people and in virtue of an implied contract with them. What we may call the leading principle of political ethics was then practically established, that the good of the people over whom it rules is the sole justification of the existence of government. Upon this point there need be no difference of opinion, nor yet as to the fact that Parliament acquired new dignity and importance as the guardian of the interests of the nation. Historians, however, though rightly convinced that the rising produced some constitutional change more momentous than can be explained merely by the transference of the crown from James II. to William III., are nevertheless mistaken in their view of the actual form assumed by what was undoubtedly a rearrangement of political forces. They hold that the revolution of 1688 transferred the power of the crown from the king to the House of Commons. It is this last opinion which while apparently true is in reality open to the most serious dispute. The reasons which tend to invalidate such a belief will be concluded only with the conclusion of the present work. But it is essential to the comprehension of the theory that the main outline of the argument on this subject should be indicated at once.

The revolution of 1688 is one of the great crises in history of which it may be confidently asserted that the full meaning has not yet been justly appreciated. Imbued with a belief in the supreme value of representative government and parliamentary power, which is regarded as the one source of political activity and national welfare, the world in general overlooks a striking feature of the alteration which was at this time effected in the distribution of political power. The increased authority of Parliament, the increased capacity of the people for interference in matters of government, is not the only point of deep constitutional interest, though popular vanity endorses the opinion of historians in this respect. There was another remarkable occurrence which passes almost unnoticed, but which confirms the theory of the present work in a very striking manner, the fact, namely, that the office subsequently called that of Prime Minister became invested with a power and importance which it had never before possessed. From the point of view of the present theory the virtual recognition of the right of the ablest available man to conduct the affairs of State is the essential feature of the revolution of 1688. It is generally stated that ministers became the servants of Parliament rather than the servants of the king, and the reader is thus left to suppose that Parliament in its corporate capacity became the ruler of the nation. Does not the real truth consist in this, that the final decision in matters of national importance which formerly belonged to the king, now passes, or begins to pass, to the Prime Minister and his Cabinet? Apart from all constitutional theories, and notwithstanding the fact that the monarch still retained a formidable capacity for interference, is it not clear that the Prime Minister rather than the king began to direct the policy of the State, while the action of Parliament in this respect remained much what it was before? The statement that power was transferred from the king to the House of Commons would seem to imply that the powers and functions of the House of Commons were modified in some decisive way, if it began to exercise powers of government which formerly belonged to the monarch, and if those were now its servants who were formerly the servants of the king. But nothing of the sort took place with regard to the House of Commons as a parliamentary body. Parliament certainly became bolder and more self-assertive, but otherwise developed no fresh and striking characteristics such as would certainly be expected if the popular theory were correct. It remained in all essential features exactly what it was before—a body which controls and modifies the action of the directing power, but which does not itself constitute that directing power. What did happen was, not that Parliament itself assumed new functions, but that out of the body called Parliament there arose a figure which virtually usurped the place of the king as the source and inspiration of government. The most distinguished member of the so-called legislative assembly henceforth assumed a position of authority over the rest, and began to exercise that influence over Parliament which was formerly exercised by the king. As a direct consequence more effective governmental management was at once apparent, and a brighter aspect was assumed by the political fortunes of England from this moment onward.

An immediate change took place in the vitality of the administration, and immediate quickening of the political pulse of the nation was perceptible, for which the growing boldness of the House of Commons, or any change in its functions as a governing body, was in no way responsible.

The reader will doubtless have anticipated the view that is here taken. England became great, and the greatness of England coincided with the advent of government by natural selection. The vital feature of the change which took place when ministers ceased in all but name to be the king's servants was not the increased power of Parliament but the increased energy and intelligence of a naturally selected Government. A king was no longer permitted personally to attempt a task which only the highest statesmanship could adequately perform, or allowed by open favouritism or injudicious preference to force unworthy ministers on the nation, and invest them with a power which their own qualifications could never in the least have justified. From this time forward, with one fatal exception, government was no longer conducted according to the views of a monarch of possibly dishonest nature, and almost certainly limited mental power, but by the intelligence of men who were in a political sense the very pick of the nation. The control of the national destinies was now committed not to the haphazard qualities of the mind of an hereditary king, but to statesmen who had attained their high position after an ordeal so severe that positive inefficiency could rarely reach office, and if it did occasionally emerge, it could not long survive the attacks of eager rivals. The essential feature of the change is thus to be discovered, not so much in the relative position of king and Parliament, as in that of the king and his chief minister. Gradually from this time onwards it is the ability of the chief minister rather than the mere word of the king which has an increasingly decisive influence. The operation of this new method of government is at first to some degree obscured by the fact that William III. happened to be fully qualified to conduct the business of the State and to be his own Prime Minister. In the reign of Anne, however, the transference of power was practically recognized, and on occasion definitely enforced. From this point onwards the power of the first minister kept steadily increasing, and thirty years later was so well established in fact that one of the accusations brought against Walpole in 1741 was that "he had usurped the sole power of directing all public affairs and recommending to all public posts, honours and employments," a power which the younger Pitt subsequently claimed as a right.

Here, indeed, was a revolution fraught with momentous consequences for the whole human race, and its effects become immediately apparent. "Once again," says a French historian, speaking of our triumphs in the great struggle between

England and France for maritime and colonial supremacy, "had England conquered, and had conquered purely by the superiority of its government." That superiority lay, not so much in any accession to the moral dignity or constitutional strength of Parliament as in the adoption of a system which raised the intellectual standard of the executive, which almost consciously recognized the inherent right of political ability to direct the national policy, and which accorded wise statesmanship an authority it never before possessed. Merely to enumerate the names of those who from 1688 take a conspicuous and decisive part in politics is to draw attention to a phenomenon of peculiar importance, and one which strongly indicates the appearance of some unique political force. Sunderland, Somers, Godolphin, Marlborough, Oxford, Bolingbroke, Nottingham, Townshend, Walpole, Carteret, Pelham, Chatham, Pitt, constitute a succession of statesmen who display in the conduct of government a continuous vigour, a frequent talent, and an occasional genius, unequalled since the days of Rome. The same spirit of emulation and activity which infused a new intelligence into the directing faculties of the State also by a kind of subtle sympathy exercised its freshening influence upon other fields of national enterprise, and revealed in all directions individual capacity which would otherwise have remained undiscovered, in India creating great commanders and administrators out of clerks and commercial adventurers, and in Canada adding a new nation to the realm. Such a phenomenon the mere manipulation of parliamentary or constitutional forms is quite inadequate to explain. It was due to the fact that a powerful stimulus was applied to national efficiency by the free play of that motive which has in all ages been the most potent cause of great political and military action. It was the simple consequence of the truth that the State was henceforth free to utilize in the service of the Government and of the nation the powers of its greatest men. This is the real meaning of the revolution of 1688.

Unfortunately though the true path of political evolution had thus been discovered, the course of progress so happily initiated was not destined to proceed without further interruption. The revival of royal power, and the renewal of the royal capacity for mischief, which followed upon the accession of George III., affords striking evidence of the terrible misfortunes which may ensue, when the management of the affairs of State lies at the mercy of a self-opinionated and incompetent monarch, rather than under the direction of the ablest of that monarch's subjects. Had England been able to adhere consistently to the policy inaugurated at the revolution of 1688, the future history of the empire might have reached an unimagined grandeur. But the old conception of monarchy as the power which had the inherent right of directing policy, though practically abolished at the revolution, still retained its hold over the minds of the people, and was once more brought into active operation on the accession of George III. The lesson of the "glorious times" of Louis XIV., that the constitutional claim of honest but obstinate stupidity to wield political power will produce consequences even more ruinous than those entailed by the calculated iniquities of a vicious ruler, is once more plainly repeated in the disastrous opening period of this ill-omened reign. Once more we see that the fatal error of absolute monarchs is not their interference with the liberties of the people, but the assumption of powers and the attempted performance of duties which their intellectual capacities do not justify. "George's main bent was to assert his individuality and take the chief share in the governance of the country." Such an aim looks harmless enough or might even seem to reflect some credit on the man who refused to play the part of a merely idle king. But the blighting influence of a catastrophe which had the effect of taking the control of national policy from one of the greatest statesmen of all time, and committing it to the mercy of a third-rate political intelligence, has been felt throughout succeeding generations, and occasionally threatens deadly mischief even at the present day. The resignation which was practically the dismissal of the elder Pitt, the premature conclusion of the glorious Seven Years' War, without any attempt to reap the full advantage that the country had won under the capable direction of an inspiring leader, were the immediate but not the most disastrous results of this pernicious return to the old theory of kingly government. The loss of the American colonies was directly due to the assumption by an hereditary and undiscerning monarch of the powers of direction which should belong to discerning and intelligent statesmanship. It is sufficiently established at the present day that the colonists had suffered no tyranny in the real sense of the term. All that was required to deal with the situation was management, one of the first qualities of statesmanship. Had the government remained in the hands of the national leaders of the English people, under the system which George III. had abolished, sufficient influence would have been given to men like Chatham and Burke to enable them to humour the discontented colonists in a way that might have arrested the crisis. And even granting that this may be a matter of speculation, it is at least certain that under a capable administration we should never have retired defeated from the struggle, and the greatest catastrophe in the history of the British empire, perhaps in the history of civilization itself, would never have taken place. This startling contrast between the course of English history before and after 1760 does not merely prove the particular fact that a great leader may produce great results: it also proves the general truth that the free career in politics is the only sure way to produce great leaders. The achievements of the seven years' war should not be regarded as the temporary result of the accidental presence of a great statesman, but as a proof of the permanent superiority of the system which brings great statesmen into prominence. It may, in fact, be fairly claimed that the difference between the glorious triumphs of the nation under the elder Pitt and its subsequent humiliation under George III. is the measure of the difference between the presence and absence of government by natural selection.

But in a country already habituated to the free career in politics the rise of individual greatness could not be permanently prevented. After this fatal interlude power was again placed in the hands of a man who was thoroughly qualified to wield it, and the country was gradually raised from the depth of international degradation into which it sank when the king undertook to be his own minister. The genius of the younger Pitt is celebrated for many reasons, but perhaps the most essential feature of his whole career has not been duly emphasized. He reduced George III. from the position of an absolute monarch in the sense now under discussion to that of a constitutional king. The King met his master. The helm was taken from the incompetent hands that had claimed it by hereditary right, and the unfaltering determination of the new pilot eventually weathered the fiercest storm that had ever burst on British shores. Thus from the revolution of 1688 the varying fortunes of the English people exactly and faithfully correspond to the character of the government; and the character of the government of the nation is decided by the amount of freedom which is given to the law of natural selection in politics. When the ability of a statesman rather than the hereditary right of a sovereign was the determining factor in the national administration, England made its mark on the history of the world. When, on the other hand, the right to direct the course of politics was resumed by an incompetent king, the nation sank under accumulated disaster. When, finally, George III. slowly began to recognize that genius alone can confer an undisputable title to political power, and was induced to assign his rightful place to a youth who was also one of the greatest statesmen of all time, the nation immediately gave evidence of a renewed vitality, and once more manifesting all its former greatness reversed the contemptuous verdict which, after the disaster of Yorktown, the outside world had prematurely deemed irrevocable.

We are now in a position to judge how far the success of Sparta may be attributed to the character and personality of the members of her Government. We find from English history that the principal advantage of constitutional monarchy consists, not in the advent of certain undefined political conditions, vaguely and inaccurately described as "liberty," but in the definite fact that the absurd pretensions of royalty no longer prevent the ablest men from assuming the direction of the affairs of State. The conditions of a limited monarchy are, in fact, extremely favourable to governmental competence, and under such circumstances the most remarkable national triumphs may be secured, even in spite of the fact that the people may not have been, in the hackneyed modern phrase, admitted to a full share in the government of the country. It is usual to decry the Spartan constitution as deficient in the essential characteristics of freedom, whatever they may be. It is certainly true that the Spartans stopped short of the disastrous licence of the Athenians, but they at least enjoyed sufficient freedom to realize what has just been shown to be its most valuable practical outcome—government by ability. As this truth, however, has never been clearly perceived, and as the advantages of constitutional monarchy have been held to consist merely in an abundance of individual liberty, and in an unlimited capacity for parliamentary interference, one of the main political causes of the greatness of Sparta has never been duly appreciated. Sparta, we are told, had two kings, *nomine magis quam imperio*: they ruled but they did not altogether govern. They seem, in fact, to have been chiefly restricted to the ideal purpose of inspiring enthusiasm and ensuring political stability. From the point of view of the present argument the essential feature of the Spartan constitution was this, that the right to direct the affairs of the State was not derived from the authority of the king but was conferred by individual ability. If we disregard unessential discrepancies in the English and Spartan constitutions, we find the general political features to be curiously similar. What was the actual amount of influence wielded by the Spartan kings it would perhaps be hard to say. But the main fact is certain, that power passed to the hands of the Ephors, just as it has passed into the hands of the English Cabinet, and with similarly beneficial results. In the Ephors we have a cabinet of five strong, determined statesmen, who owed their position, like English cabinet ministers, in theory to the vote of an assembly, in reality to a system of selection according to ability, just as the English Cabinet are chosen. For though it is an assumption, it is an assumption justified by all we know of the Spartan State, that the post of Ephor was filled by strong and dominant personalities, who though they owed their position to election, owed it at the same time to a natural pre-eminence in valuable political qualities, and who thus fulfilled the requirements of the present theory, as will be shown in a succeeding chapter. Oman tells us, "The Ephors could practically return whomsoever they chose to act as their successors in the ensuing year," and this renders it reasonably certain that ability to forward the interest of the State was the sole ground of choice. It is true that we do not meet with many distinguished names in Spartan political history; but fame was largely won in ancient times by doubling the parts of statesman and commander-in-chief: and though two of the Ephors accompanied the king on military expeditions, to secure conformity to the general policy of the State, they undertook no actual military command. While the military efficiency of the Spartans was clearly due to this wise separation of military from political duties, a great source of renown or reputation was at the same time closed to their leading statesmen. We may conclude, then, that the success of Sparta was directly due to the increased efficiency of government which, as we see from English history, is the natural consequence of the institution of limited monarchy; an efficiency which may be the less hampered in producing great national results, because of the very absence of those popular conditions of "freedom" and "self-government" which are supposed to contain the whole secret of political success. And though it is true that the necessary subordination of the interests of the individual to the welfare of the community was carried to a point where it began to defeat its own object, yet Sparta under the Ephors became one of the most famous States of the world, because a system of natural political selection gave to her public services an efficiency which was the secret of her brief but immortal career.

The theory that the success or failure of a nation is intimately connected with the permission or the prohibition of the free access of talent to the ranks of the administration, receives a further illustration in the history of Spain. The very idea that genius confers a natural title to political supremacy would have appeared as the rankest blasphemy to this king-intoxicated nation. An administration of almost incredible incapacity was the natural result of the belief that no great intellectual requirements are implied in the work of successful government, or that if they are, a divinely appointed king must of necessity be capable of fulfilling them. A blind adoration of the person and office of monarchy, almost without parallel since the days of the Persian empire, precluded the very possibility of the idea that the admission of talent from the body of the Spanish nation could in any way contribute to the efficiency of the Government. The supreme control of the nation was not a human, hardly even an earthly office. It was a divinely ordained ceremonial, and its performance belonged to the king alone, whose counsels were directed by God himself. This royal monopoly of the intellectual functions of government was in itself enough to involve the Spanish empire in consequences as fatal as those which followed a similarly exclusive policy in France. But when, in addition, we are confronted with the overweening power of ecclesiastics who demanded unquestioning obedience to their decrees, and who further discouraged the employment of intellect in politics by practically forbidding the exercise of the mental faculties except for the purposes of saving the soul, we have a code of political ethics almost ideally designed for the ruin of a devout and noble-minded nation. When we consider the character of the Spanish people, as it was displayed in the individual achievements of her most brilliant period, a startling contrast is to be observed between the appalling ineptitude of the Government and the amazing vigour and active genius of the nation as a whole. Whatever may be the verdict on the doings of the Spaniards from the moral point of view, they exhibited at one period an enterprise and endurance unsurpassed in the history of the world. "The history of Spain in the seventeenth century," says Buckle, "is the history of one long and uninterrupted success. That country . . . was able to annex the whole of Portugal, Navarre, Rousillon . . . she acquired Artois, Franche Comte and the Netherlands: also the Milanese, Naples and Sicily and Sardinia, the Balearic Islands and the Canaries," and a page follows of further successes at home and abroad. These successes, it will be observed, were almost entirely military, because the art and conduct of war was the only national enterprise upon which individual talent was encouraged to exert its powers. The achievements of the Spaniards in war, and the superhuman exertions which their colonial pioneers exhibited in the acquisition of empire, remain to this day the wonder of the world, and amply prove that the individuals who composed the nation had all, and more than all, the qualifications which, properly directed, go to build up permanent, not ephemeral greatness. The failure was not in the individuals: their powers were, on the contrary, exceptionally brilliant; the inference is therefore irresistible that the failure was in the Government; and the failure was in the Government because the intellectual life-blood of the nation was not permitted to reach and reinforce the centre of administration. This view is confirmed by the fact that the only effective national energy which Spain displayed for more than two hundred years was under two strangers of genius, Alberoni and Ripperda. But the leadership which was thus borrowed from two brilliant foreigners, and which arrested for a time the disastrous decline of the Spanish fortunes, might well have been continually supplied by the Spanish people themselves, had not the effects of disuse finally caused an atrophy of native political talent. The truth of the present theory, and of the fatal effect which absolute monarchy may exert by the exclusion of intellectually superior men from government, might almost be proved from a study of the reign of Philip II. alone. "During more than forty years," says Mr Ellis Barker, "that ill-starred monarch Philip the Second had ruled the gigantic Spanish world empire, and his government had proved fatal to it." Yet if from the time of Charles V. the passionate desire for distinction, so characteristic of the Spaniards, had been made to depend for its satisfaction upon service done to the community, the history of Europe would have been very different. Don Quixote remains as the eternal symbol of the waste of individual enthusiasm which, practically utilized in the government of the country, might have raised the Spanish empire to more than Roman greatness. But the greatness of Spain was ephemeral, because this brilliant individual genius, though allowed to perform isolated services in war and colonial enterprise, was not permitted to permeate the Government, and thus re-act with beneficial influence upon the character and fortunes of the nation as a whole. The permanent greatness of a nation, in spite of all theories to the contrary, depends upon the greatness and intelligence of the Government, and no Government can continue to show such greatness and intelligence unless its energies are perpetually reinforced from the most capable part of the nation. In the case of the Spanish, as in that of the French, a magnificent stream of ability, enough and far more than enough to ensure the success and prosperity of either nation, was diverted from its proper course, and allowed to run to hopeless waste. The Spanish Government, deprived of its natural sustenance, literally perished of inanition, and the failure of its mental powers produced a general paralysis, which spread insidiously through every branch of national enterprise.

Those who lay more than a merely conventional emphasis on the value of government are generally thought to be advocating something which undermines the manliness and self-reliance of a nation. The importance attributed to government in these pages is, in fact, in marked contrast to the general opinion of Englishmen, who are frequently taught to take a pride in believing that in their past history they owe everything to themselves and nothing to their rulers. In his once celebrated " History of Civilization " Buckle tries to show that the failure of France and Spain was due to the decay of individual initiative caused by the over-carefulness of their rulers. English governments, on the contrary, according to theorists of this kind, did very little for the people: they left difficulties largely to solve themselves, and we are asked to believe that this alleged careless and irresponsible attitude has been of inestimable advantage to the English character. The fallacy of this reasoning is becoming more and more apparent at the present day. Because the course of action adopted by the rulers of France and Spain was ruinous to either country, this merely proves the obvious conclusion that a bad government may produce ruinous effects. In the opinion of Buckle, Government is largely superfluous because it merely does for people what they ought to do for themselves. This, however, is an entirely erroneous assumption based on a complete misinterpretation of the relation of the individual to the community. Good government does not hinder the development of individuality, but assists it. The individuals of a nation cannot realize their full possibilities except with the aid of able government. The alternative to the irrational government of the French and Spanish was not the absence of government which Buckle seems to demand, but the presence of a better and more rational government, consisting of the ablest men that the nation could produce. Good government does not replace individual initiative: it stimulates and at the same time organizes private effort, performing for the nation as a whole functions which it cannot possibly perform for itself. No nation, however self-reliant and self-sufficient the individuals of which it is composed may seem to themselves to be, can in reality dispense with intelligent guidance. Contempt for this truth has been and is still the ruin of republics and democracies. Such intelligent guidance knows not merely when to intervene, but also when to abstain from intervention. It implies the conception by the ablest intellects of a policy by which the national energies may develop their utmost activity, and attain their highest expression. The alternative to Buckle's chaotic individualism is the careful organization of national enterprise and effort under a wise and watchful Government; and the time is rapidly approaching when only an effective combination of this kind will save the empire from the dangers which confront it. Again, there is a subtle power which may perhaps be called the psychological influence of government, and which must be taken into serious account in estimating the forces which make a nation strong and great. According to this view the chief characteristic of good government is not legislation but leadership, the leadership which can evoke individual enthusiasm and direct it in the highest interests of the community: and this power of leadership is a quality which neither constitutional government nor representative institutions are able to provide without simultaneously calling in the aid of the principle of the present theory. Finally, good government implies a faculty which is beyond all price, and which is at the same time almost purely intellectual, the faculty, namely, of making the right decision in an emergency. In the life of every nation there is not merely one period but an incessantly recurring series of periods when a decision must be taken which vitally affects the future of the country. What that decision may be depends upon the character of the statesmen who are in power. One mistaken step in a great emergency has before now ruined a nation: a series of such mistakes is certain to do so. That nation is most successful which makes the fewest mistakes in politics: and the fewest mistakes in politics will be made by the community whose political system encourages the supremacy of the ablest of its sons.

We may conclude, then, that at a certain stage of political evolution constitutional progress consists in excluding the hereditary monarch from the exercise of a predominant influence on national policy, with the result that the business of government falls into the hands of those best fitted for its conduct. The surpassing service of monarchy in the general evolution of government will be considered in a succeeding volume. The character of that service is somewhat changed under constitutional monarchy, yet the uses which are left are still of inestimable importance. The subordination of individual ambition to the highest interests of the country could not have been so effectively attained in English history except for the simultaneous existence of a spirit of self-devotion inspired by the influence of the crown. Constitutional monarchy has the merit of inspiring enthusiasm for the service of a sovereign who is the emblem of imperial unity, while at the same time individual capability becomes the determining factor in the choice of those who direct the policy of the State.

CHAPTER III

PARLIAMENT AND THE PREOCCUPATION OF HISTORIANS

IF the conclusions of the preceding chapter are correct, a new idea must be introduced into the study of history. Hitherto it has been taken for granted that the most important feature in the history of government is the growth of parliamentary institutions. The evolution of government to all intents and purposes is supposed to be identical with the evolution of the political assembly. It will be shown in the following pages that this belief in Parliament as the sole source of the active powers of government is mistaken, and that governmental forces having quite a different origin demand equal consideration. The attention of historians, for reasons to be presently assigned, has in fact been unduly concentrated upon one feature of the evolution of government, to the exclusion of another, equally permanent and indispensable. No elaboration of parliamentary functions can alter the immense importance of the human and personal factor in government, the character, namely, of those who suggest and individually carry out the policy of the State.

It is impossible to get away from the fact that the fortunes of a nation depend in the last resort upon the capabilities of its leading statesmen. According to the view which will be maintained in the succeeding pages, Parliament supplies only the regulating, moralizing, and controlling force of government. But the intelligence, activity, and vitality of an administration are derived, not from Parliament, but from particular individuals, who owe their position to the process of political selection described in the previous chapter. It is indeed true that the rise of the ablest to political power is a process from which the State cannot derive the fullest advantage except in connexion with the simultaneous development of the political assembly. But the necessity of first proving the main thesis of the present work must postpone to a subsequent volume the full discussion of the relative parts played by the leaders of the nation on the one hand and of the political assembly on the other. So far as the subject is treated in the present work, the argument will be restricted to disputing the belief in the omnipotence of Parliament and to showing what the political assembly does not do rather than what it does.

The evolution of government is usually taken to consist solely in one movement, but it consists as a matter of fact in two. We have, on the one hand, a tendency of which the result is to place selected ability at the service of the State: we have, on the other hand, the gradual assertion of a power of control on the part of the people, which ensures that this political talent shall be exercised in the highest interests of the community. Constitutional development cannot therefore be properly apprehended unless, besides the growth of parliamentary institutions, the historian simultaneously takes into account the conditions which affect the influence of great individual statesmen. The rise of the politically ablest and the development of the political assembly are correlated phenomena in the history of government, and their reciprocal action constitutes the double movement from which a healthy political evolution proceeds. The argument, therefore, of the present work resolves itself into the contention that in addition to the process by which Parliament with its controlling powers is developed, we have another process by which, in accordance with natural law, government is incessantly supplied with those qualities of intellect and character which are indispensable for national success.

But in order that this process may take effect, the political initiative belonging to the king must be replaced by the political initiative of a leading statesman. To limit the intellectual rights of monarchy is a procedure as necessary, and has an effect as important, as to abolish the moral pretensions of absolutism. Only when the king relinquishes his claim to the intellectual leadership of the government of the nation does this distinction come within the reach of able men who could not before aspire, and tend to pass to an individual of adequate mental powers. Yet the vital significance of this change is overlooked by historians in consequence of their preoccupation with moral and constitutional considerations of seemingly greater importance.

If we consult the ordinary text-books in order to discover what is the process by which in English history government has been rendered a more efficient instrument of national welfare, we find that the assertion by Parliament of " the liberties of the people " is supposed to sum up the whole theory of governmental progress. The attention of the reader is almost wholly directed to the struggle by which the community establishes over the king a parliamentary control, designed to abolish arbitrary power and to enforce the truth that the end of government is the good of the nation. The assertion by parliament of its rightful moral powers has unquestionably been necessary to prevent kings from ruling in defiance of the first principles of good government. But it is not sufficiently recognized that this object may be successfully achieved, and that the affairs of the kingdom may yet be very seriously mismanaged. The action by which Parliament prevents the king from governing for his own selfish ends, immensely important though it is, is not the sole requisite of national welfare. Those immediate interests of the realm which do not involve any great general principles, but which may none the less concern the very existence of the nation, are more closely identified with the ability and personality of the leading statesmen than with the question of the royal deference to Parliament. The eye of the observer should, accordingly, be fixed not merely on the struggle of Parliament with the king, but on the facilities offered for the rise of leading men. The intellectual powers of an administration, its capacity for taking the wisest course during some grave political crisis, and for successfully applying national means to great national ends, are subjects which demand as much consideration as the prevention of unjust government on the part of the hereditary monarch. The accomplishment of these duties does not belong to parliamentary action as such, but to individual statesmanship. Parliament may successfully insist upon government being just, but it cannot call into existence the intellectual characteristics which make government great. The individual mind of the great statesman is not a sort of by-product of the parliamentary mind, as we are frequently encouraged to believe, but something with capabilities and self-determined energies of its own. The rise of political ability capable of dealing with the immense difficulties, both external and internal, which beset every administration, is therefore of equal importance with the development of the powers of Parliament. It has already been shown that in the revolution of 1688 the free access of talent to office was a result as significant as the well-known parliamentary features of this historic struggle. It follows that the interest of constitutional history must be extended so as to include, not merely the dual relation between king and Parliament, but the triple relation between king, leading statesmen and Parliament. The defeat by Parliament of the king's claim to govern as he pleases is important; but equally important is the gradual rejection of the royal right to be the sole director of the policy of the State. A sound governmental evolution therefore depends not merely upon forcing the king to assent to a proper system or a reasonable theory of government, but also upon allowing political ability to make good its claims to political leadership, since, as already pointed out, the intellectual pretensions of the king retard political progress no less than his moral pretensions. There are, in fact, two great results which naturally follow from the struggle with absolute monarchy, namely, government on the best general principles and government by the best available men; yet of these only the first has attracted general observation. The king's attitude to Parliament is the sole object of constitutional interest. In reality, however, the king's attitude towards political talent is of equal and occasionally of even greater moment, since it is unfortunately within his power to neutralize the suggestions of a statesmanship which might otherwise have carried a nation to the highest point of greatness. This truth has been incidentally pointed out in the previous chapter. But a glance at the reign of

James I. will serve to illustrate more fully the present contention, that the establishment of even the most effective parliamentary control over absolute monarchy constitutes only the half of what is in reality a double phenomenon. The greatest parliamentary triumph ever achieved would, in fact, be largely barren of permanent and decisive results, if it did not have the effect of opening a career for political talent, as well as of correcting the abuses of arbitrary power. If it is in any way possible to estimate the relative necessity of these two requirements, government by ability, as stated in the previous chapter, is the more indispensable, since self-preservation is, in the last analysis, the preliminary condition of national as of individual existence.

The interest of the reign of James I. is supposed to centre in the struggle of Parliament against the King's claim to dispose almost at will of the lives and fortunes of his subjects. It is, of course, true that the welfare of a country cannot be assured except on the condition that the king acknowledges his constitutional dependence upon Parliament: it is essential for future political progress that the king should be forced to admit that he does not rule in his own right, but merely for the good of the nation. Yet the most triumphant vindication of the supremacy of Parliament over the crown merely indicates the proper general theory upon which government should be conducted, but does not ensure the ability necessary for its immediate success. In every reign, however, there are in reality two governmental questions which affect the welfare of the country—one is a question mainly of the future, the other is the question of the present; one is the theory upon which government is to be conducted, while the other is the character and personality of the actual members of the administration. The personality of those who conduct the government of a country is obviously, with regard to contemporary issues and contemporary management, a consideration of even greater importance than the theory on which it is conducted. A perverse selection of ministers by the king may in a few years irreparably injure or disastrously affect the international position of the State. Consequently, the fate of the country may for the moment depend, not upon the king's attitude to Parliament, but on the amount of influence which he permits talented statesmen to exercise upon the conduct of affairs. To insist on the power of arbitrary taxation, to deny the authority of Parliament on this and kindred subjects, is doubtless a form of procedure which can on no account be tolerated by the community. But to rob the country of the services of its greatest statesmen by disallowing their appointment or refusing their advice is equally criminal, since it denies in fact, if not in theory, that the welfare of the nation is the true object of government. The question, therefore, in the reign of James I., and in every similar reign, is not merely whether the king will duly recognize the authority of Parliament as the guardian of the rights of the community, but whether his complacent self-sufficiency or impenetrable conceit will lead him to withhold from better men the opportunity to make or keep their country great. Throughout the reign of James, after the death of Cecil, great statesmen and great men of action were simply not permitted to exist ; and the fact that this prohibition of the very possibility of greatness was followed by a painful deterioration in the moral and intellectual atmosphere of the nation appears, to the present writer at least, to constitute a greater crime against humanity than his attitude to Parliament. Yet historians, who are aroused to indignation over any slight offered to Parliament, dismiss the annihilation of the political intellect of the country with a sarcastic comment, or a sigh over the occasionally perverse consequences of hereditary monarchy. The king of course is, from the point of view of the constitutional historians, acting within his theoretical rights. But they fail to see that if this is so, the theory of government which permits such action is as worthy of condemnation as the most deliberate contempt of Parliament. To assert the supremacy of Parliament is to affirm the general principle of good government ; but at the same time to leave the king in control of the policy of the State is to omit the particular measure necessary to carry the general principle into effect, and is productive of consequences more immediately ruinous to the interests of the nation. A comparison of the reign of James with that of Queen Elizabeth will help to make this clear.

The greatness of the reign of Queen Elizabeth was no doubt largely due to her power of evoking enthusiasm, and to her capacity for eliciting the noblest efforts of her subjects in the promotion of national greatness. But for the successful administration of the affairs of a troubled kingdom, in a troubled and turbulent period of the world's history, something more is needed than the power to evoke enthusiasm. That additional requirement Elizabeth was enabled to supply. In addition to her personal advantages, and her unquestionable affection for her people, she possessed another priceless virtue of a sovereign—the capacity to recognize the ability of those of her subjects who were intellectually greater than herself, and the moral courage to submit herself to their political guidance. By her own greatness of character, by her own instinctive sense of what was due to her people, Elizabeth grasped a truth which two revolutions and the whole of the intervening course of history failed to teach George III., that when the interests of a nation are at stake, superior statesmanship in a subject must at any cost to kingly dignity be duly recognized and honoured. None of her failings can mar the glory of having divined the righteousness of a constitutional theory which some contemporary sovereigns have not learned even at the present day. Her vanity may have been inordinate, her parsimony contemptible. Her only reason for action may have been truly feminine, the impossibility of further hesitation. But in spite of all this she exhibited the highest statesmanship which is possible for a sovereign, namely, the willingness in the last resort to stand aside in favour of a talented subject. To James, on the contrary, the greatness of others was offensive, and he endeavoured in consequence to retain the whole management of the kingdom in his blundering, incompetent hands, with the result that he fell into subservience to the Spanish ambassador, or into bondage to worthless favourites. The degradation of his reign has been felt throughout successive generations, and Englishmen have felt that degradation, not because James endeavoured to thwart the will of Parliament, but because genius and enthusiasm withered in his presence, and all possibility of political greatness disappeared under the malign influence of his offensive personality.

Historians frequently write as if the duty of a reigning monarch was not so much to secure the happiness and welfare of his own actual subjects as to supply lawyers of future generations with satisfactory precedents for the elaboration of their constitutional theories. Disobedience to modern principles of democratic government is, in their opinion, a more serious offence than the annihilation of political talent. For most of them there is only one really serious offence, disregard of the authority of Parliament, although those qualifications of a king which are most pleasing to this type of theorists are not necessarily those which make an efficient ruler. The recognition of constitutional limitations by the king is insufficient unless accompanied by a readiness to confer political authority on the ablest of his subjects. If he fails to encourage the activity or to utilize the assistance of the greatest statesmen of the realm, no amount of subservience to Parliament will atone for the disasters which a preference for the exercise of his own autocratic but limited intelligence will bring upon the helpless country.

Finally, if the political influence of individual talent is of supreme national importance, the attitude of Parliament towards great statesmen demands as careful consideration as the attitude of the king. However great the importance of Parliament may be in other respects, its abstention from all unnecessary interference with the individual initiative of its leading members is one of the most indispensable conditions upon which the effective working of government depends. When a political assembly, as not infrequently happens, is so jealous of greatness that it will not entrust distinguished individuals with the powers necessary for efficient administration, or when for any other reason it usurps those functions which should belong to leading statesmen, such an attempt at collective management is followed by ill-considered political action both at home and abroad. The faculty of actual government does not in reality belong to a collective body, and the erroneous belief that a Senate or Parliament in its corporate capacity has ever successfully administered any kingdom or empire will be subsequently exposed. When assemblies endeavour to perform in earnest the task of administration loosely ascribed to them by historians, they overstep the limits of their true political usefulness, and enter on a course of action as detrimental to the highest interests of the State as was the autocracy of Louis XIV. or George III. Freedom of action for statesmen, within the limits prescribed by a rational constitution, is the policy which produces the best national results, and the undue restriction by Parliament of the rights of individual genius and individual initiative is a contingency as undesirable as the appointment of unworthy ministers by the irresponsible caprice of royalty.

The facts, therefore, which have been set forth in this and in the preceding chapter seem to point to the conclusion that there is in existence a law of nature tending to furnish government with qualities which are indispensable, but which parliamentary action does not supply. In proof of this contention it has been shown that one of the chief results of the constitutional struggle against absolute power has been to allow the intellectual guidance of the State to fall into the hands of competent statesmen, a remarkable phenomenon which has entirely escaped observation. One cause of this remarkable oversight on the part of historians has already been suggested. It will, however, be conducive to a readier acceptance of the present theory if the reasons for this strange indifference to one of the most important facts in the evolution of government should be formally investigated.

(1) In the first place, the history of science shows that a law of nature may be in active existence without at the same time being obvious: its importance seems in no way to hasten its revelation. The length of time, therefore, during which some great principle of evolution has been ignored constitutes no argument against its possible validity. The limitations of the human mind are such that the most far-reaching truths must remain undiscovered, until the necessary preparation for their discovery is in all respects complete, and in the case of social science this condition has not long been fulfilled. Meanwhile there are many vital processes, both in bodily and national life, indispensable to a vigorous existence, which have for a long time acted without recognition, silently discriminating in favour of those who unconsciously fulfilled the requisite conditions. In this category is included the need of intellect in government, which, though largely unperceived, has not on that account been the less imperative. If in the scheme of nature the satisfaction of even a vital want had been dependent upon the perception of that want, the progress of the human race would long ago have been arrested. The intellectual requisite in political life which man virtually ignores, nature supplies,

proportionately rewarding those communities whose constitutional arrangements have ensured a constant supply of talent to the directing faculties of the State.

(2) In the second place, the perception of a new truth is rendered difficult in proportion to the firmness with which an older interpretation is already established: and the firmness of the belief that the political assembly is the sole source of political progress has precluded the appreciation of the existence of other most important governmental forces. Political assemblies are associated with the first traces of government consciously undertaken for the good of the people as a whole. Parliament, as we shall presently see, has been solely responsible for preventing the incessant abuse of governmental power, and has thus protected the highest moral interests of the community; and for this reason the welfare of the nation seems to be inseparably and solely identified with the growth and history of this most justly celebrated body. The lines on which the historical investigation of governmental problems should proceed have thus been predetermined, and the eyes of historians have, for an apparently sufficient reason, been turned in one direction alone. Nothing of interest, nothing that is intimately concerned with the successful administration of government, has been expected outside the phenomenon of the political assembly, and therefore nothing has been found. The full appreciation of truth, like the

appreciation of beauty, depends on the predisposition of the observer's mind, and the significance of even the most important facts may be entirely missed if prevailing theories are so framed as to dispense with the necessity of their existence.

(3) In the third place, reformers and historians have been so fully occupied in considering the moral deficiencies of government, that the dangers of inadequate intelligence have almost entirely escaped observation. Human nature is so constituted that the question of the honesty and benevolence of an administration has throughout history entirely overshadowed the question of its intellectual competence. Though to govern wisely is, on a scientific analysis, not less desirable than to govern well, yet because the injustice of a government touches the life, the liberties and the self-respect of the individual, it is habitually resented with a bitterness which mere stupidity has only occasionally aroused. One of the most striking and persistent features in the history of every people is the systematic evasion by government of those very duties which theoretically it is bound to perform. At least one half of the political records of the human race are occupied in relating the desperate efforts made by each progressive nation to teach the governing power its true relation to the community. Though in constitutional theory there is no possible answer to the assertion that the end of government is the welfare of the nation, yet in historical fact there is

at first no recognition whatever of this apparently self-evident proposition. To inspire government with a true conception of its duties, and to force it to obey elementary ideas of justice is one of the first conditions of progress. To Gibbon the Antonines seemed the only honest rulers in history, nor has the course of politics since his time been such as to render this conclusion entirely out of date. It is no wonder, therefore, that in the absorbing attempt to secure a reform in the moral attitude of government, the intellectual requirements of the situation should have been entirely left out of sight. If a person begins by denying that he is bound by the ordinary laws of morality, and has in consequence no duties to perform, it is idle to discuss his ability to perform them. It has never been possible for any active nation to relax for a moment the effort to induce a better frame of mind in their rulers. Let the reader turn over the pages of such a work as Hallam's "Middle Ages." He will quickly be made to feel that the momentous issues depending throughout Europe upon the desperate struggle against political injustice, are to this earnest writer so absorbing as to preclude the possibility of studying government from any other than a moral point of view. Government is first and foremost a matter of deep practical interest, and until the great practical question of the moral supremacy of Parliament, and all that it involves, is placed beyond the possibility of dispute, the human mind does not feel sufficiently at ease to deal with the merely scientific, intellectual, and speculative aspects of the evolution of a political control. Yet the necessity of securing intelligence of administration is not the less imperative because of the more obvious necessity of securing justice. However important or indispensable it is for a nation to encourage reforming action on the part of parliament, it is not the sole condition of political progress. The struggle of the political assembly for what is called "liberty," however successful it may be, is insufficient to ensure national well-being unless it is attended by conditions which render possible the access of ability to positions of political influence. The requirements of intellect cannot be left out of sight, and in our gratitude to reformers and to Parliament we must not be led to underrate the value of the merely talented statesmen. The action of the Pyms and the Hampdens is necessarily more thrilling, and makes in consequence a greater appeal to the dramatic instincts of mankind: but the success of a nation depends upon other requirements besides the protest against political wrong. It depends upon the conception and execution of great political designs, upon the skilful guidance of foreign and domestic policy, upon cool and cautious wisdom rather than on fervent moral protest or heroic self-devotion. Clear-headed statesmanship, whatever the motive by which it is inspired, is the permanent need of a nation; unselfish reforming fervour only an occasional necessity.

(4) The belief that the injustice of government is the sole cause of the unhappiness or misery of nations carries with it the further conclusion that this defect can be remedied by parliamentary action alone. The attention of the world has therefore been concentrated upon the reform of parliament as the sole requisite of political success. In modern times this view has found its expression in a constitutional movement, which though effecting an object as necessary and valuable as the destruction of absolute power, is nevertheless in itself quite unable to supply all the requirements of good government. The new political panacea which is supposed to obviate the necessity of individual ability is the modern representative system. When Parliament even after its triumph at the revolution of 1688 still failed to give complete satisfaction, this was attributed to the fact that it was not sufficiently representative: and up to a certain point this reading of the political situation was correct, for there was work to be done which intelligence alone would have been quite unable to perform. Even after the right divine to govern wrong was no longer claimed as a royal prerogative, the moral education of the nation's rulers was still very far from complete. The misrule of the king merely gave place to the injustice arising from the pressure exercised upon the administration by an aristocratic House of Commons; and the remedy correctly adopted was identity of interest between government and governed, secured by the principle

of representation. It is, however, characteristic of all political innovations that they give rise to exaggerated hopes on the part of their promoters. It was the confident assumption of the school of political thought associated with the names of Austin, Godwin and Bentham, that the attainment of ideal administration was merely a matter of sympathy between government and governed, and that a Parliament truly representative, and therefore incapable of passing unjust laws or imposing unequal burdens, would solve the final secret of political success. Since the days of Bentham, therefore, the main object of certain reformers has been to bring government into complete conformity with the wishes of the people; and the idea is still prevalent at the present day that nothing else is necessary to secure the perfection of the State. Yet the assumption that the sympathy between government and governed, which representation tends to promote, is alone sufficient to secure social and political well-being, is entirely groundless, because though it may cure the moral, it leaves the intellectual deficiencies of government untouched. In the prevailing conception of politics it would in fact seem to be assumed that so long as the hearts of statesmen are sound, the qualities of the head are a matter of indifference. It would in truth be an easier and a happier world if the wish to do good implied also the ability to do it. Yet there could hardly be a greater political heresy than to regard excellence of intention as the measure of administrative ability.

The value of representation will be fully admitted in a succeeding chapter. But the limitations of this as of every other political device must be clearly recognized. It is the great achievement of representation to correct class selfishness, and to do this is to remove one great cause of administrative failure and national misery; but to correct class selfishness is not, as many believe, to solve the whole problem of government. Again, representation may or may not lead to correct diagnosis of political trouble, but it most emphatically does not imply the certainty of cure. The very common assertion that a proper system of representation would have remedied the evils from which France was suffering from the time of Louis XIV., and that representative meetings of the states-general would have prevented the French Revolution, is made in complete disregard of these essential truths. According to this view it is not the absence of government by ability, as stated in the last chapter, but the absence of government by representation, which was the cause of all the trouble. In such a case, however, it is obvious that representation in itself can do no more than draw attention to the ills of an afflicted nation. Representation is indeed necessary in the first place to enable the political sufferer to detail the symptoms of his malady, and in the second place to prevent government evading its just obligations;

but the cure depends upon the cleverness of the treatment prescribed by the political physician. Representation is, in fact, an effective remedy for the sufferings of a nation only if it is accompanied by or gives rise to a statesmanlike ability to remove them. This, however, is in essence government by ability, not government by representation. Exclude from the idea of representative government all extraneous notions of intellectual eminence, and its champions are bound to assert that the mere statement of grave political and social difficulties must necessarily lead to their cure, a proposition which is rendered more than doubtful by our modern knowledge of the extreme complexity of the social organism and of the disorders from which it suffers. Again, it must be pointed out that however valuable its influence upon internal affairs, representation affects only the domestic policy of a nation, and is in no way concerned with the provision of the trained powers of intellect absolutely necessary for the successful adjustment of its foreign relations. Enthusiasts can maintain their confidence in this principle as a political panacea only by ignoring one-half of the forces that affect the life of the State, namely, those which act upon it from outside. They seem never to admit the existence of incessant external dangers which representation is entirely powerless to meet or avert. In this respect, a representative system would have been no help to France, since it can be safely asserted that even if France succumbed to internal difficulties, those difficulties were themselves largely due in the last analysis to a defective foreign policy, upon which representation would have had little or no effect. Representative institutions help to make government just, but they are not designed to make government wise; and the central defect of the administration of Louis XIV. was not so much its injustice as its unwisdom. It may be freely admitted that to restore France to a perfectly sound condition, to effect a permanent cure of the miseries of the lower orders, and to prevent a thoroughly unjust division of the weight of the national burdens, a controlling power representative of the whole nation would have been needed in addition to government by the ablest. But the point is that in all such cases the presence of ability is the initial requisite, and parliamentary representation only a supplementary precaution. Granting the necessity of the dual arrangement, statesmanlike ability is the more immediately indispensable, and this forms no part of the theory of representation, though as a matter of fact it is a not unusual accompaniment. The beneficial results which have been found to follow representative government are in reality due not merely to representation, but to the fact that among the representatives is frequently found a man of sufficient ability and strength of character to grapple intelligently with the difficulties of the situation. Unless the introduction of a representative system into France had permitted or encouraged the predominance of the politically fittest, a contingency most improbable under the jealous eyes of Louis XIV., it would have revealed the presence of complicated disorders, but it would not have been able to trace those disorders to their real cause, or to effect a real cure : it would doubtless have prevented the unjust incidence of taxation, and might possibly have curtailed the personal extravagance of the King ; but unless the right to control the national policy had been taken from his hands and entrusted to those of competent statesmen, the real source and origin of the impoverishment of the French people would have remained as before. If to the merits of the representative system are added the effects of government by ability also, the universal belief in its efficacy is justified, but not otherwise. This is what has happened in England. The fame of this political device is largely due to the fact that it is universally credited with triumphs which belong to government by ability alone. Nor is this exaggerated belief in the all-healing virtue of representation the only case, as we shall presently see, in which achievements due to the sheer intelligence of statesmanship are ascribed to more popular and impressive causes.

(5) In the fifth place, the supposed uncertainty of the appearance of talent in a political administration has generated a disbelief in the possibility of influencing the intellectual character of government. If a phenomenon is regarded as spasmodic,

uncertain, incapable of control or rational prediction, it loses interest for an age which is preeminently scientific. Accordingly one of the most important phenomena in the study of politics is still regarded as belonging to the region of chance, and is believed to possess characteristics so elusive that they could not be regarded as a possible subject for scientific investigation. A recent writer in an able work on government and politics speaks of "the personalities of the statesmen whom chance has brought to leading positions." This opinion is absolutely typical of the prevailing belief. The appearance of great talent, or even the presence of more than ordinary intelligence in government, is everywhere regarded as an entirely fortuitous occurrence, for which indeed as the occasional gift of the gods we may be thankful, but of which it is impossible to take account in the study of national evolution. It is this universally prevailing assumption which has prevented the perception of the steady influence exercised by the desire for patriotic distinction in bringing talented men to the head of self-governing states. The rise of great men being thus regarded as merely accidental, not only is their permanent value underrated in consequence, but the influence which they have exercised at certain particular crises of the world's history is to a large extent unperceived. Accordingly, some of the greatest national triumphs which can be clearly traced to the inspiring and directing genius of great political leaders, are not attributed to their proper cause, the intelligence of the government, but to certain moral qualities of the governed, because this interpretation is thought to reflect more credit on the character of the human race. Of these moral influences which are supposed to explain all national greatness the idea of liberty is by far the most popular, and this is the more unfortunate since the idea of liberty is entirely lacking in that scientific precision which is as desirable in history as it is in any other form of investigation. Of this determined moral bias a remarkable instance will now be given.

(6) Finally, then, it may be asserted that the value of intellect in government is not systematically recognized, because its effects are attributed to moral causes, and more especially to the idea of liberty. The vague and almost superstitious veneration which is paid to the idea of " liberty " establishes a species of mental prepossession which prevents any accurate estimate of the political process actually taking place at a given time, which forbids the ascription of effects to their real causes, and withholds full recognition from the achievements of statesmanlike ability. Preoccupied with the belief that the realization of a " free government " is the one cause of national success, historians create a mental atmosphere in which they readily see their own anticipations justified. Any prosperity which may follow the adoption of " free institutions " is regarded as the

necessary consequence of the moral elevation which accompanies the realization of this beneficent change.

In the first place, however, it is to be remarked that the enthusiasts of freedom manage to ignore the significance of the undoubted fact that liberty does not always lead to satisfactory results, but may under certain circumstances be productive of moral and political disaster. Obviously, therefore, it is not liberty alone which can be credited with invariably healthy consequences, but liberty combined with certain other advantages, and it remains to be seen what those other advantages are. In the second place, it is not perceived that the conditions which attend "free government" are also the conditions which render possible government by merit, because a free career is opened to political talent. Liberty may or may not be what it is usually called, the breeding-ground of virtue: that depends on the attendant circumstances: but it infallibly discloses the presence of political ability should any exist. The almost invariable rise of great men in times of national crisis or revolution, or during periods of new-found freedom, is simply an illustration of the tendency of the principle of government by natural selection to find fulfilment whenever artificial obstacles are removed. Yet this phenomenon, though well known, has not been traced to its real cause: the unwonted facilities offered to the able and ambitious under the circumstances described constitute an entirely new factor which is never taken into accurate account ; and national results which in reality belong to talented guidance and inspiring leadership are in consequence ascribed in a vague and general way to the unique magic of liberty and freedom. Conversely, when under absolute government the absence of political selection has lowered the intellectual quality of the national leaders, and political mismanagement has produced its inevitable consequences, the national ruin which ensues is attributed to the moral degradation which besets the character of the people when they are excluded from the management of their own affairs. The deficiencies of this method of reasoning with regard to the history of England after 1688 have already been pointed out, and it has been shown that our success was due to the fact that the leading minds of the nation were released from intellectual subservience to the king, and that the counsels of wise statesmen were henceforth permitted to have something like their full effect. It will now be shown that another great triumph usually ascribed to "liberty" and "free institutions," the victory of the Greeks over the Persians, may with greater accuracy and justice be assigned to the effects of superior political wisdom and more intelligent general leadership.

The usual explanation of this great event is that the moral advantages enjoyed by the free self-governing Hellenic States rendered them unconquerable by the slavish myriads of an oriental despot. The increased energy and initiative imparted to each individual Greek by the consciousness of his political heritage is supposed to have inspired an irresistible impulse which brought victory in its train. It need not be denied that the consciousness of freedom has under certain definite conditions a valuable moral effect, and it is of course obvious that to give all the citizens a personal interest in the prosperity of the State is to increase the chances of national success. Profit-sharing as an incentive to exertion and individual initiative is undoubtedly as valuable in politics as elsewhere. But to make these admissions is in no way to diminish the force of the present argument. Liberty, except with certain definite or implied restrictions, does more harm than good, while the profit-sharing principle alone is as utterly insufficient to account for the success of a great nation as it is to account for the success of a great business. In politics as in commerce it is, in fact, practically useless without clever and successful management. The freedom of Athenian life for a time favoured the accession of genius to political management, and then, and only then, did their national affairs prosper. When, however, they used the perfect freedom of democratic choice to place incompetent rulers in control of the policy of the State, we are enabled to perceive from the ruin which ensued that the freedom of the people to choose their rulers is valuable only on the condition that it contributes to the political supremacy of intellect. The admission of the ordinary political shareholders to a part in the management of the national estate is a procedure attended with the most extreme danger, and success in such a case as that which we are discussing depends not upon the sympathetic accuracy with which government represents the wishes of the people, nor yet upon the liberty to indulge in a choice of leaders without regard to the real interests of the State, but upon the existence among the individuals of intelligence and self-restraint sufficient to induce them to use their freedom for the purpose of entrusting the conduct of national affairs to the ablest of their fellow-citizens. During the Persian wars these requisite pre-conditions were fulfilled. The free political system of the Greeks had the effect, at this time and on this particular occasion, of placing them under the leadership of some of the most remarkable men that the world has ever seen. In the person of Themistocles and the other Greek leaders, intellect had assumed that national supremacy which it always assumes under favourable political conditions, and had produced its natural effect upon the situation. When we leave vague generalities and endeavour accurately to trace effects to causes, it seems unquestionable that the Greeks owed their success mainly to superior statesmanship and superior military science. It was the better organization and direction of national effort which gave them the victory, and this was the result, not of liberty, but of intelligence. Intelligence and efficiency, and not merely the moral exaltation which arises from political freedom, though this of course was present, were the decisive factors in this momentous contest. Free government at this period of Greek history, but not afterwards, meant government by the fittest: it meant the inestimable advantage to be derived from the almost superhuman wisdom which government by natural selection had brought to the aid of the Athenian State. On the other hand, despotic government meant in the case of the Persians the very opposite to all this. It meant that they were bound to follow the stupid and suicidal plan of campaign suggested by the overweening vanity of a typical Eastern potentate, and were compelled to yield obedience to a conception of statesmanship and strategy such as might have been expected from the half-demented megalo-maniac whom the accident of birth had seated on the Persian throne. Never did fate set two nations to play the deadly international game with a more monstrous handicap than this. The preliminary conception of the whole campaign was so utterly unintelligent and futile that the Persians, had they only known it, were beaten even before they crossed the Hellespont: they were defeated at that early stage, not because they were slaves while the Greeks were free, but because they were captained by folly, vanity, and irresolution, and the Greeks by genius and clear-headed determination. In the case of the Persians the sole direction of all matters of national concern, the sole choice of military and political subordinates, and the sole direction of the strategy of the campaign, was left in the hands of one of the most incompetent leaders of that or any other time. The political system of the Greeks, on the other hand, allowed genius in war and government to assume their proper place, and so to give a triumphant direction to the fortunes of the whole community. The immense numbers of the host of Xerxes which, to the uninstructed imagination, seem to enhance the difficulty and the brilliance of the victory of the Greeks, as a matter of fact rendered it much easier. These very numbers were to the Persians a disadvantage of exactly that kind which is entailed by sheer stupidity of leadership, since the maintenance and efficient tactical direction of this unwieldy mass would have overtaxed the powers of a Hannibal or a Napoleon. The whole campaign was in fact marked by a series of strategical and tactical blunders sufficient to bring about the defeat of the Persians many times over, without in any way taking into account the moral characteristics of the opposing nations. Had the mobility of the Persian forces not been destroyed by enlargement to gigantic and unwieldy proportions, had their efficiency not been impaired by the inclusion of worthless elements for the sake of mere show, had proper strategic positions been occupied, and tactical impossibilities such as the attack on the famous pass of Thermopylæ, been left unattempted, in accordance with the most elementary principles of the art of war, in a word, had the leadership employed on the Persian side been anything but a compound of royal vanity, cowardice and folly, the issue of the expedition would hardly have been doubtful. In all human probability the patriotic, devoted, and united Persians would have been victorious over the separate and selfish communities of the disunited Greeks, and in that case the world-famous demonstration of the superiority of Greek liberty to Persian slavery would not have taken place. The victory of the Greeks was intellectual rather than moral: they won not so much because the individual soldier was free, but because his leaders were wise: and the Greeks had wise leaders because, at the time of the Persian invasion, their liberty was of the nobler kind which encourages the influence of the great and wise statesmen whom it calls into political activity. Thus one of the most conspicuous triumphs in all the history of free government was in reality a triumph of intellect, which had established its right to national leadership in exact accordance with the requirements of the present theory.

In conclusion the general argument of the present and of the preceding chapter may be summed up as follows: The history of the evolution of government does not consist merely in the development of the political assembly. Besides all the great advantages which are brought about through the agency of the political assembly, history gives evidence of the existence and beneficent activity of another phenomenon equally important, the rise of the talented individual to political power. The success of nations has been in proportion to the facilities offered to men of intellect for engaging in the work of government. The leaders who are employed in the working of a constitution are as important as the nature of the constitution itself: the intellectual quality of those who conduct the government deserves as much consideration as the theory under which it is conducted. The character and personality of leading statesmen is not a matter of accident, as is generally supposed, but depends upon a law or tendency of nature, which has the effect of inducing the relatively fittest to come forward in the service of their country. To prevent undue interference with the action of this law is therefore a matter of not less importance than the assertion of the rights of Parliament, since discouragement or prohibition of the exercise of political talent is sufficient to cause the arrested development of the nation. This view of the evolution of government is also confirmed by what is known as the "history of freedom." Although historians have bestowed an especially careful study on the history of "freedom," only one of its results has succeeded in attracting their attention. It not only limits absolute authority, emphasizes the right of popular control, and ensures the triumph of the principle that government exists only for the good of the people, but it brings into operation political talent which would otherwise remain hidden. It reveals the truth that there are two tendencies in the evolution of government, one of which encourages the rise of political ability to a position of predominant influence, while the other develops or strengthens the controlling assembly. The political greatness of England, for instance, is due, not as is usually believed to parliamentary institutions alone, but to government by the ablest under parliamentary institutions. Finally, it may be asserted that the constitution of human society is such that a community can only put forward its best powers, and realize its highest expression, when the intellectual qualities of its government are of the highest order. And the satisfaction of this requirement is under the care of a law of nature quite distinct in its operation from that which promotes the growth and development of the political assembly.

CHAPTER IV

ELECTION AND REPRESENTATION

IT is now necessary to enquire what is the relation of the present theory to a fact of the greatest importance in the history of politics, the phenomenon of election. In the opinion of many people, perhaps of most, government is merely a matter of election. Stripped of unnecessary detail the whole process would seem to conists in appointing suitable persons for the purpose of devising and carrying into effect the regulations necessary for harmonious social existence. Though the whole matter is not quite so simple as this, the fact remains that the practice of electing those who are to occupy the chief positions in the State is a feature of political life apparently in sharp contradiction with the theory of government by natural selection. There would seem to be at first sight no common ground between government as the result of the free choice of the community and government as the result of a natural law which automatically determines the question of political supremacy.

This difficulty, though for some years it appeared to the present writer insuperable, is in reality no difficulty at all. Not only does it disappear when the idea of election is subjected to analysis, but the result of the investigation is to confirm in a remarkable manner the two fundamental principles upon which the present theory of political evolution is based. It is indeed characteristic of the confusion which, after some 2500 years of discussion, still envelopes the subject of government, that the vaguest notions exist as to what election is and does, and as to the part which it plays in the political life of the nation. If, for instance, we accept the explanation that the people choose those to govern them whom they prefer, the question immediately arises, what are the grounds of this preference? When these are given, the scientific enquirer is in the next place not only entitled but impelled to ask: Are the grounds of preference justified by the highest interests of the State? This in turn compels an investigation of the object of the State, and of the ends to which political action should be directed. Finally, when this all-important problem has been determined, it still remains open to discussion whether popular election is the method best adapted for securing those ends. It appears, therefore, on investigation that this apparently simple question of election is intimately connected with the deepest problems, not merely of government, but of life itself.

Without, however, insisting that no one has a right to give an opinion on the elective method without first having solved the riddle of existence, it is easy in another way to show that the difficulties raised by the problem of election are very far indeed from being properly apprehended. At the very outset of the investigation we find that there are two aspects of election, between which no distinction is made in popular thought, and which even advanced thinkers are wont to confuse together, but which are in reality quite separate phenomena. Election in popular phraseology is taken to be synonymous with representation, and the identity seems to be proved by the very general adoption of the term "elected representative." Representation does indeed take place by means of election, but it is something quite distinct. This is proved in other ways, but particularly by the fact that election existed thousands of years before the idea of representation was even so much as thought of. The process of election, in the sense in which it will be presently explained, is as old as the rudimentary forms of political organization. The idea of representation, on the other hand, is of quite modern origin, and was practically unknown to the human race before the eleventh century. Election belongs to the earliest, representation to the latest phase of political evolution. Election, therefore, would seem to be a thing intimately connected with the original and fundamental principles upon which government rests, while representation satisfies a want which arises only at an advanced stage of human development.

If, then, leaving the question of representation to be investigated later, we can discover the root idea of election, it will help towards the comprehension of the fundamental principles of government. The attempt, therefore, will first be made to analyse the meaning of the original process of election, such as it was before the idea of representation arose from the growing complexity of modern political life.

The truth is that there is no contradiction between the principle of election and the theory of the present work. Election and the survival of the politically fittest are closely connected and even complimentary phenomena, for they are different phases of the same process at different stages of political evolution. Election is the recognition of superiority.

In the Homeric poems and other records of the political customs of antiquity, we hear of councils of elders or assemblies of the tribe or nation electing a king or ruler. If we wish to analyse the meaning of the phenomenon of election, we must first discover for instance what were the motives which induced the Franks to raise a certain person as king upon their shields: why should such an assembly choose one person rather than another? Or why, in a Homeric gathering, would an Agamemnon have a better chance of election than a Thersites? Obviously because the human qualities of Agamemnon, his birth, intelligence, appearance and reputation, give the promise of more effective leadership from him than from Thersites. In the case of the Franks, the hereditary antecedents, martial attainments, impressive personal qualities of their prospective king, his power of command or the glory of his achievements, inspire an instinctive recognition of his superiority, clearly mark him out in their opinion as the worthiest object of national honour, and justify them in their determination to select him for a ruler. It is plain, therefore, that the considerations which recommend a leader are the qualities calculated to secure the good of the community, and the process of election is robbed of all national meaning unless this is the object at which the assembly aims. Accordingly the conclusion seems unavoidable that election, so far as it is rational, is the recognition of a superiority qualifying a certain person for the successful prosecution of the duties of a ruler. But it is the essence of the present theory that this result is also attained by the operation of natural selection in government. The question therefore arises whether the difference in the process by which this identical result is attained is so great as to negative the possibility of an evolutional harmony between the two methods. Is there, in fact, any necessary contradiction between the idea of government as the result of conscious political selection, and government as the result of natural selection acting through the struggle for distinction?

When political assemblies determine on the

election of a king or a leader, they are not endowed, either individually or collectively, with any magical faculty which will enable them to dispense with ordinary powers of observation, so as to be able to detect the presence of ability or genius by some wonderful process of intuition. The very wisest assembly that ever has existed or will exist is compelled to rely upon such evidence of capacity as can be gathered from the general record of the character and achievements of the candidates. And when we say this, we are saying nothing else than that the competition of life has already marked the leader in advance, before the assembly has set the seal of recognition upon his merits. Though he may now be a leader by virtue of the will of the community, he was all the time actually or potentially a leader by virtue of natural selection. The will of the people confers upon Agamemnon no quality of leadership which he did not before possess, nor can the utmost exercise of its capacity make good the deficiencies of Thersites. We may, therefore, at once answer the question as to whether there is any necessary contradiction between government as the result of the elective principle, and government as the result of natural selection in politics. There is no contradiction, in spite of the difference of procedure, for the simple reason that the elective principle must be guided by the verdict of natural selection before it can make a rational choice. When Themistocles, Pericles, Scipio Aemilianus, Chatham, Pitt, are described as having been the freely chosen leaders of a self-governing community, the real meaning of this is obvious. What is loosely called election in the case of these leading statesmen is in fact nothing more than acquiesence in a political precedence already virtually achieved. It is clear that their stronger character, more impressive personality, more striking political or aristocratic antecedents were the considerations which gave them an influence which general consent merely confirms. The rules of evidence upon which the elective principle must proceed are borrowed from the process of natural political selection, and conform to a conclusion already established by the facts of life.

Many people, however, will be strongly inclined to dissent from the opinion that the will of the people as manifested by election is simply restricted to giving an assent to a conclusion already established by the course of events. It will be urged that such a view of the matter does not take into consideration either the right of self-government or the free choice of the community. With regard to the right of self-government, it will be shown in the final chapter that in its real meaning it contains no idea opposed to the recognition of political supremacy. The presumptive right and supposed ability of the electors to exercise a completely unfettered choice will now be examined.

The first point to be remarked with regard to the right of free choice is the curious mental stipulation which invariably accompanies the demand. It is habitually assumed by those who regard such an exercise in "moral freedom" as essential to the well-being of the community, that the people will utilize their liberty for the purpose of making a good choice. License to ruin the State by a deliberate preference of bad rulers to good is never for a moment contemplated by any sane-minded statesman. If freedom were the real point, it would be a matter of perfect indifference to constitutional theorists whether the rulers chosen were good or bad, and the universal ruin of the political communities of the world would afford a lively satisfaction to the logical metaphysician, provided only it were the result of the free choice of the electors. But since only those who are in bondage to the tyranny of this idea of liberty are prepared to assent to so logical a conclusion, it is evident from the political and practical point of view that the essence of the matter is not the freedom but the excellence of the choice. Regarding the question historically it is easy to explain the strangely mistaken emphasis which has been laid upon the freedom when in reality it is the goodness of the selection that is meant. It has been the almost universal experience of mankind that the worst evils of misgovernment have been inflicted by arbitrary rulers in whose appointment the people have had no share, and the natural inference has been that inability to choose their own type of administration has been the sole cause of the grave political injustice beneath which the world has so often groaned. By an instinctive act of reasoning the conclusion has been reached that since the people desire their own political happiness, control over the selection of rulers is the best way to secure it. The assumption is that their wishes and their interests are identical, and that therefore the concession of freedom must by the ordinary laws of human nature lead straight to the desired end—political happiness. The right of free and independent election is thus seen to be of practical value only on the assumption that it leads to improved governmental relations: it is obviously merely a means to the end—political well-being. Free choice merely means the absence of any form of compulsion which would interfere with the natural and presumptive tendency to choose what is best. If, however, this natural and presumptive tendency should happen to be absent, all the practical advantages which have inspired the belief in its supreme virtue instantly and inevitably disappear. Freedom of choice is a political privilege which brings ruin in its train unless it is a wise choice: it is in fact strictly conditioned by the national necessity of securing excellence of administration. Election, even if we insist on qualifying it by the word free, still means election of the best, and therefore remains as originally defined, the recognition of superiority.

The same point can be shown in another way. If we compare the position of a director of a financial company with that of a leading statesman, and enquire what are the reasons which justify election in either case, the parallel will be found instructive. Let us imagine a situation where the shareholders of a company choose their director just as the electorate choose their ruler. The director is the "creation of the will" of the investors in this financial undertaking, and his salary is paid in consequence of the belief that he is the best manager available under the circumstances. The shareholders have thus in a certain sense been enabled to exercise freedom of choice. Yet the capabilities of the director whom they choose, and not the fact that they are free agents, is the point of importance to the company. It might, indeed, seem quite unnecessary to remind a person of ordinary intelligence that the director of a company is chosen for the performance of certain duties requiring a more than ordinary degree of ability. The continued existence of the company as a profit-making concern somewhat obviously depends upon the success of the shareholders in making a good selection. Yet this fact is liable to be overlooked if the question is the precisely analogous one of the choice of a ruler by the community. In the case of a democracy the advocates of "free institutions" insist that the moral right of the democracy to "please themselves" is the supreme consideration, and that if this is duly secured, the happiness of the community must inevitably follow. Would the shareholders of a company, would any persons, discussing the affairs of a great commercial undertaking, think even for a moment of being content with this aspect of the case? It is quite certain that they would not, and the reason why they would not is equally clear. They would know that the real point was not whether the director was the choice of the shareholders, but whether he was qualified, by marked financial power, to administer the affairs of the company in a thoroughly satisfactory manner. Unless he were able to improve or maintain the position of the company as a paying investment, the question whether he was agreeable to the shareholders, or whether he was the free choice of these prospective money-makers, would be absolutely and entirely irrelevant. What, then, is the clear and unmistakable implication? It is this, that the choice of the shareholders, their inclination in favour of one director or another, is absolutely without rational meaning unless it has regard to the question of competence. We may call them free and independent shareholders if we like, just as political voters are called free and independent electors. But the shareholders are not really free. Freedom in one respect, and that the most important, cannot be predicated of them. As rational beings who wish to increase their dividends they are on the contrary bound to choose as a director the individual who is most likely to raise the value of the company's shares in the money market. If some fanatic were to endeavour to persuade a meeting of shareholders that the real point was to exercise their choice as free individuals, regardless of all financial consequences, he would certainly end by being summarily ejected from the room. If shareholders choose the right man the company succeeds, but it succeeds not because their choice was "free" but because it was rational, and was therefore directed upon a proper and fitting person. If the old quibble is brought forward that freedom and wisdom are identical, and that the electors are free because a free choice is necessarily a rational choice, such a contention concedes the point at issue, that efficiency is the decisive consideration. The whole question turns upon the power of the shareholders to recognize talent in a manager, and the necessities of political and commercial administration in this respect are absolutely identical.

The conclusion, then, to which the previous argument points is as follows: The object of the existence of a financial company is the pecuniary welfare of the shareholders. It is essentially a profit-making concern. But a State is also nothing else than a profit-making concern, in respect of the fact that its members desire their own welfare, and are in the habit of meeting together to take measures for the purpose of securing this result by the choice of a fit administrator. This "free and independent choice" is therefore strictly conditioned by the rational obligation to choose a director who will raise the value of political stock, the profit in this case being the increased security, happiness and welfare which will accrue to the individual from the results of good government. The election of statesmen is not practised by the civilized world as an exercise in moral freedom, but from the conviction that such a method is better calculated to secure the general interests of the State. The only concession that can be made to the educational idea is that a nation must have practice in the art of choosing for the purpose of being enabled to make a good choice. In order to train the political insight of the people, upon which in democratic times so much depends, they must be exercised in the use of their political judgment, even at the risk of occasionally making mistakes. If this practice is also calculated to strengthen the national character, such an advantage is incidental; the real initial object remains precisely what it was before. Freedom of election is in the first instance only valuable in proportion as it conduces to superiority of government. Whatever moral benefits electors may derive from their independence, they are never released from the inexorable conditions on which alone political existence can be successfully conducted: their supposed freedom is found to be identical with the logical necessity of choosing the fittest character or the highest ability at their disposal. Where freedom of election has been successful, it has produced its good results not because the people had the rulers they preferred, but because the grounds of their preference were justified by right reason. The fatal national consequences which ensue from a neglect of this fundamental truth will be described in the examination of the elective policy of the French, Dutch, and Athenian republics given in the final chapter. It will there be shown that one of the most ruinous defects of a democracy is the tendency to elect statesmen because of qualities which may perhaps be admirable in themselves, but which have no real bearing upon the vital points at issue. From these and innumerable other instances we arrive at the inevitable conclusion, that literal insistence upon the right of perfectly free and independent election, so dear to the minds of metaphysical historians and visionary political thinkers, means in practice the ruin of the State.

We are now in a position to appreciate a fact of much importance to the general theory of the present work. The phenomenon of political election may be regarded from two points of view. If we are in a mood to fix our eyes upon the personality of the statesman, upon the history of his life, and upon the successive steps in his career by which he has achieved political eminence, we are accustomed to speak of his rise to power. The use of this phrase indicates with perfect justice that from one point of view he has been the creator of his own fortunes. If, however, we are thinking of the constitutional aspect of the case, we forget or ignore his active contributions to his own triumph, and describe his rise to political power as being due to the nomination of Parliament or the wish of the electors as registered in the ballot box. This double aspect of election is not only admitted at different times, but is recognized and popularly described by a single expression. In the ordinary political terminology of the present day we speak for instance of a candidate for Parliament "winning an election." In this phrase the double character of the elective process, as a rivalry between individuals on the one hand and as an election by the people on the other is very clearly apparent. If, indeed, we were to insist upon analysis of the ideas contained in this expression, it is evident that they come very near to a contradiction in terms. Strictly speaking, it would seem clear that a man cannot win a position by his own efforts, and yet at the same time have it given to him by another set of persons. Yet as a matter of fact this is precisely what does happen, and the phrase escapes condemnation because the apparent confusion of terms corresponds to a real confusion of processes. A successful candidate for a seat in Parliament is in reality engaged in two operations which correspond to the double relation in which he stands, first to the other candidates, and secondly to the electorate; for he both wins from the point of view of the rivals whom he defeats, and he is elected from the point of view of the majority who cast their votes in his favour. We are thus enabled, in a popular but none the less significant way, to observe the actual co-operation and harmony between the struggle for distinction and the principle of election; and to the definition of election as the recognition of superiority we may now add that it is also the method by which the rivalry of individuals is made to serve the uses of the community. The victory of one or other of the rival candidates and the choice of the community are found on analysis to be identical. Election is an individual success as well as an act of national choice, and this fact, rightly regarded, is in itself sufficient to place the present contention absolutely beyond dispute.

As a practical illustration of this point we may take the contest (or election) for the presidency of the United States. It is not a little significant that in what is supposed to be the most democratic of nations, the presidential election should take the form of an undisguised, and at the same time a somewhat unscrupulous and undignified struggle for supremacy between two or three individuals. The sonorous platitudes in the exordium of "The Constitution of the United States" display a lofty disregard for the phenomenon of individual rivalry, yet it remains the fundamental feature of a constitution which is perhaps the most elective in the world. The presidential election, like all other elections, decides between rival personalities, and the supreme people acquiesce in the situation as the most natural thing in the world. Such an apparently irrelevant contest ought, from the point of view of constitutional principles, to seem entirely out of place. It is rather regarded, in spite of political theories, with evident appreciation and delight, as something quite appropriate to the quadrennial inauguration of government. The high-sounding phraseology of constitutional documents may ignore this element of human rivalry: but it remains, and will for ever remain, a permanent feature of all political evolution.

It is now possible to offer an answer to the question which was propounded at the beginning of the present chapter, namely, the relation between the principle of government by natural selection and government as the result of the "free" choice of the community. As we have just seen, there is, in a relatively advanced community, a double process at work. On the one hand the ruler or statesman rises to distinction and political power through his own exertions. On the other hand the principle of election acknowledges his power and influence as indicative of administrative qualities which tend to increase the efficiency and importance of the State. If electors do not make this consideration the basis of their choice, election is undistinguishable from a meaningless or pernicious exercise of arbitrary authority. The true function of an electorate or of an elective assembly is to conform to the law of natural selection in government, or in other words to follow the verdict that has been registered by the ordeal of political life. The elective process is therefore subsidiary to the law of natural selection in government, for the simple reason that it depends for any good results it may produce upon the previous decisions of that law. States have often been successfully ruled by a leader who has come to the front by his own exertions in the struggle for power, but never in the whole history of the world by rulers elected without any regard to the proper qualifications for power. From the evolutional point of view election, instead of being antagonistic to the struggle for supremacy, is seen to be in reality nothing more than a supplementary process. It does not create capacity, it simply certifies its existence and confirms its authority when found. The conclusions at which we have arrived may therefore be summed up as follows. There is only one rational object of election—the discovery and utilization of political talent for the good of the nation as a whole. The electorate is the human medium through which political selection takes place. The free choice of the community, though it may increase the sympathy between Government and governed, introduces no essential change into the fundamental conditions which decide the health and happiness of a State, but merely reaffirms, under the forms of modern constitutional procedure, the already existing truth, that the best available men are likely to form the best available administration. Election frankly recognizes, as we have seen, the struggle for distinction, but gives to individual ambition a social meaning, by making political rewards conditional on national service. It remedies the inevitable shortcomings of this natural process, making more accurate and effective that selection of able men which under the original system must remain liable to a varying degree of injustice and uncertainty. Political election is merely the verdict of natural selection delivered (theoretically with more precision) by the voice of the people. It is the recognition and adoption of a superiority which is due to natural causes, and which exists quite independently of any elective process.

The function of election in recognizing, utilizing, and rewarding political merit has been shown. The meaning of that much later birth of political evolution, the process of representation, is now to be considered.

Representation may be regarded as having two forms. There is the representation of geographical areas, and there is the representation of opinion. By the first it gives to increased numbers of the population, and to extended areas of the kingdom, a direct influence on the government which without the aid of this principle they could not possibly exert. In the second place, it influences the character of government by sending members to represent certain lines of political theory and to impose certain general rules of political conduct. This last form may be called the representation of opinion. Whether all representation is not in the last analysis a form of the representation of opinion is a question which will be reserved.

Representation has played an important part in the history of political evolution. It is, in fact, the device which alone has rendered possible an indefinite increase in the size of a State, consistently with the preservation of sound constitutional arrangements; so great are the difficulties, though merely of a physical nature, which conditions of space would otherwise have introduced into the problem of the expansion of the area of government. Many readers of history will remember that, according to ancient ideas, a homogeneous community was strictly limited in size from the supposed necessity of assembling the whole of its members simultaneously for the purpose of consultation. Even the genius of Aristotle was only able to conceive of a State of such moderate dimensions that its members might at any time be collected within the circumference of the agora. For, as he quaintly says, if this limit be exceeded, what herald that had not a voice of fabulous proportions could succeed in making himself heard by the whole assembly. The size of the State is, in fact, in his opinion, limited by the "carry" of the human voice. From this formidable dilemma humanity was released by the simple numerical expedient of regarding one individual present as the equivalent of indefinite numbers that were not present. Though its origin is humble and obscure, though it seems, in fact, to have been at first extended, perhaps invented, merely for purposes of taxation, it soon assumes, in accordance with a well-known evolutional tendency, a wider significance. It becomes an essential feature of politics, rendering possible the government of outlying portions of a kingdom with as much attention to general local welfare as if the administration could, in Aristotelean fashion, personally address all the assembled members. It was ignorance of this principle which to a large extent was answerable for the misgovernment of ancient empires. Had this ingenious method of making one equal a thousand been known to the Romans, and had they been willing to adopt it, it is conceivable that the increased cohesion given to Italy and the empire might have hastened the advent of modern civilization by a thousand years.

The underlying idea of representation, it will thus be seen, is not that of governing but of influencing the Government. It aims at bringing the detailed interests of the citizens before the rulers of the country, and at enforcing a more sympathetic attention to wants that might otherwise be disregarded. Those portions of the community, whether districts, classes, or trades, who consider their interests to be threatened or their welfare neglected, are enabled through their representatives in Parliament to say to the Government, "We are here in person, and you cannot therefore refuse through ignorance to give us the attention we demand, or to remove the disabilities from which we suffer." It is especially to be noticed that representation affects only Parliament, or, in other words, the controlling power of government as distinguished from the Government itself. The idea of representation is not the idea instinctively associated with the activities of the Cabinet. Parliament, on the other hand, is identified with the representative idea, because it is a body whose function it is to watch the action of government in the interests of the governed, and because representation is the method by which it has been induced to perform this duty adequately, as will be clear from the following considerations.

History shows that Parliament suffers from the same defect as that which it seeks to cure—a tendency to act in its own interests alone, and to promote only the welfare of those classes with which it is connected. A political assembly is found during the earlier years of a nation's constitutional history to take its complexion entirely from those members of the State who by wealth and hereditary position have the greatest political and social influence. During the Middle Ages and up to the end of the eighteenth century, for instance, the composition of Parliament was mainly aristocratic, and the government was therefore mainly conducted in the interest of the upper classes. Representation comes into existence, or at least is used, to remedy this defect, and by widening the sympathies of Parliament widens the sympathies of the government which it controls.

The same kind of influence which Parliament exercises over government, the people by means of representation exercises over Parliament. Parliament imparts to government the tone which it received from representation, and until all classes are represented impartial government is rarely or never obtained. The guardians themselves are continually tempted to become unworthy of their trust, and representation, by removing all excuse for the selfish exercise of parliamentary powers, is the answer which the progress of political evolution has provided to the famous query of Plato, "*Quis custodiet ipsos custodes*?" Representation is thus the culminating process of a political evolution, which secures great men for the service of the State, subjects them to the control of a political assembly, and then by the principle of representation makes the political assembly itself just, sympathetic, and efficient.

It will now perhaps be seen how greatly the inner meaning of representation differs from that of election. There is, in fact, nothing really in common between the old-world principle of election and the modern principle of representation. The popular movement which modifies the constitution of Parliament so as to enlist its protective powers on behalf of all, without distinction of class, aims at something quite distinct from the original purpose of election. The object of election is to secure ability: it is engaged in deciding the question of the relative superiority of individuals. But the theory of representation makes individual merit entirely subordinate, since the individual by hypothesis is merely the embodiment of the interests of those for whom he stands: though it may reveal grievances, it is no part of its function to provide the talent necessary to cure them, nor is such a stipulation in any way inherent in the original conception. Yet though the root idea of each of these principles is comparatively simple, as much cannot be said for the combined part which they have played in history. It becomes, in fact, essential to attempt to determine the very difficult question of the relation in actual political life between the principle of election and the principle of representation.

It was pointed out at the beginning of the present chapter that considerable confusion of thought existed on the subject of election. It must now be added that, as between election and representation, there is not merely a confusion of thought but a complete confusion of practice. If we ask whether members of Parliament at the present day are chosen for personal merit, or merely as convenient instruments for representing a set of opinions, it would be hard to give a definite answer. Both ideas are combined together, with results that will be presently examined. For instance, the same individual may, on account of conspicuous personal qualities, be selected to fill an important cabinet post, and may yet at the same time be regarded as representing the interests of a certain particular portion of the community. Again, if representation is regarded as a means of procuring special governmental attention to a particular portion of the kingdom, this does not release the representative from obligations to the nation as a whole. The member for Durham is concerned with the interests of Great Britain as well as with those of his own constituency, and this the vast majority of electors are sufficiently patriotic to understand. They know that the member who represents their district is also elected for the purpose of upholding and increasing the power and dignity of the country, and regard any measures that he may take with that object as reflecting honour upon themselves—his constituents. He is their representative in name only, and in reality the representative of the country, sent up by a certain towsnhip as evidence of their participation in the political life of the nation. "Representative" is in the greatest number of cases merely a name for a person, who is locally identified with a particular centre, but who otherwise belongs to the nation. In proportion to the political distinction of a member his representative qualities seem to disappear, while his public duties towards the State take the precedence of his private duties towards his constituents. Thus the Prime Minister and the members of the Cabinet all represent certain towns or districts, but not even the most pronounced separatist would suppose that they had been elected for the purpose of giving more attention to their various electorates than to the general welfare of the nation. Whatever, then, may be the real distinction between the principle of election and the principle of representation, they are seen in practice to be hopelessly confused: and this confusion constitutes one of the most fatal defects in modern political life. The origin and the nature of the trouble therefore demands the most careful consideration.

Representation, as already pointed out, was originally a method of securing the attention of Government to localities, interests, or classes which it would otherwise have neglected. Such a method is no longer necessary. Representation as a means of securing justice for distinctive portions of the kingdom is, in fact, a complete anachronism. The theory of government in the interests of the whole nation, and of each separate portion of it, is already so firmly established, that no special machinery is required to enforce the truth that one town or district counts politically for just as much as another. Representation in its original sense is obsolete, and does not, as a matter of fact, take place: and we introduce confusion by continuing to talk as if it did. Representation at the present day means nothing but the representation of political opinion, as is shown by the history of the Reform Bill. What is remarkable about the Reform Bill, and what is of great importance to the present argument, is that though such re-distribution bills may seem to be concerned with the representation of localities, they are in reality concerned not with the representation of localities at all but with the representation of opinion. The accidental feature of the Reform Bill of 1832 was the enlargement of the electoral area so as to include large unrepresented towns like Birmingham. The real meaning was the extension of the franchise, not to certain localities, but to certain classes that had hitherto been excluded. The representation of new areas meant in reality the more effective representation of a new type of political opinion. This change, which can be traced in operation during the period of the Reform Bill, has now become definitely established. A member of Parliament is supposed to represent the interests of, say, Haslemere, Newcastle, or Hartlepool. But half an hour's observation of the passions of an election day would serve to convince the most casual spectator that if the member does represent the local interests of these places, it is in a very subordinate degree, and that he is in reality fighting for the interests of one or other of the two great political parties. The fact that he resides at Newcastle or Hartlepool, or merely has his committee rooms there, has no bearing whatever upon the real point at issue. The representation of the opinions of the Conservatives or Liberals of the whole country has completely submerged the idea of the representation of local interests. We arrive, then, at the following position. Representation has become entirely the representation of opinion, and is the means by which the democracy, through Parliament, exercise a control over government. That such a control should somehow be exercised by the people over government has been consistently admitted throughout the present work. The question which concerns us here is whether the legitimate exercise of democratic control over government could not be performed in a way which does not interfere with the fundamental conditions of sound administration. According to the theory here maintained the true object of election is to secure political ability. Representation, on the other hand, while aiming at the legitimate object of associating the nation more closely with the conduct of government, entails under present conditions an almost complete indifference to the actual political capabilities of those who direct the nation's course. The principles on which members of Parliament are elected are tending in a direction contrary to the encouragement of political genius, because the representation of political opinion is interfering with the proper function of election, the election of the best. Though this amounts to a serious, and, if continued, to a fatal defect in the theory of government, or the working of the constitution, the remedy is fortunately quite simple. The representation of opinion must take place in another way, in a way which does not interfere with the election of the politically fittest.

Let us first certify the facts. " A general election," says Bryce, " although in form a choice of particular persons as members, has now practically become an expression of popular opinion on the two or three leading measures then propounded and discussed by the party leaders, as well as a vote of confidence or no confidence in the ministry of the day."[1] Such a practice tends fatally to lower the quality of an administration by treating members, not as prospective statesmen, whose intellect and personality are of the highest importance, but as mere pawns in the political game, whose noblest duty is to outnumber their opponents, and to sweep them off the political board. The ability of a politician thus tends to be a secondary consideration, and if he is only eligible from the pecuniary point of view, and professes the right creed, his personality will serve as well as that of any one else to increase the party majority and "count two on a division"! It is imperative that the controlling influence of the electorate over government should be exercised by some means which does not degrade the individual candidate into a mere numerical unit, whose chief merit consists in his power of increasing by one the numbers of his political party.

It might indeed seem a sufficient derogation from the true purpose of government that the members of Parliament should be chosen, not so much for personal merit, as for the profession of certain opinions which are forced in lump upon them. Yet the weakness and confusion which this system

[1] "American Commonwealth," vol. i. p. 466.

tends to introduce into the government of the country is further increased by the habit of using a general election as a means of deciding the fate of a particular measure, with the result that members are cast, like living beans, into the voting urn of Parliament, in order that the views of the sovereign people may be ascertained upon some political question of the day. Democracy, it would seem, makes as contemptuous a use of its human instruments as the most typical of eastern potentates. Finally, the consequences of this evil system are not confined to the election merely of the rank and file of Parliament. They tend to influence the intellectual quality of those who hold the most important positions in the State. This was substantially the view of Mr A. L. Godkin, who pointed out that "all the great modern democracies have to contend almost for existence against the popular disposition to treat elective offices as representative, and to consider it of more importance that they should be filled with persons holding certain opinions or shades of opinion than by persons most competent to perform their duties."[1]

When a nation like England, which leads the world in political development, makes so serious a mistake, we may be sure that there is an evolutional reason for this departure from common sense. The truth is that the progress of political development has brought us to a point where a

[1] A. L. Godkin's "Unforeseen Tendencies of Democracy."

differentiation of functions, previously combined, has now become necessary. That tendency to economy which induces nature to adopt an old organ or institution to a new purpose seems to hold good in political evolution. When the political life of the English nation had reached a certain stage of development, the need of representation began to make itself felt. Accordingly the old method of election was made to serve the new need of the representation of localities, interests, and opinions. Thus it has come about that the representation of opinion by means of the election of members of Parliament disastrously interferes with the most important functions of government. An attempt is, in fact, being made to secure two distinct objects—the composition of an effective national administration and the expression of political opinion, by an identical method. At an earlier stage of evolution the adoption of an old means to express a new need may have been comparatively harmless. At a late stage it introduces incessant confusion into politics, and interferes with the efficiency of government by interfering with the proper function of election. Such a method of expressing political opinion lowers the character of an administration in the first instance, and then by placing it in dependence upon a majority vote, tends to make its tenure uncertain and its tone subservient. Members of Parliament should owe their political position to some distinction of intellect or character, or if this requirement is too ideal, they should at least not be degraded into the position of mere political counters.

From the point of view of the individual elector the position is this. Every person entitled to register a political vote is acting in a double capacity. He is called upon to perform a duty on behalf of the community, while at the same time he is permitted to exercise a personal privilege or it may be a right on behalf of himself. The duty he is called upon to perform is to assist in the election of the ablest available persons for the purpose of conducting the government of the country. The privilege he is allowed to exercise is that of giving an opinion on the general policy of the State, and occasionally also of deciding the fate of some particular and important measure. These two perfectly distinct functions he is forced under the present system to perform by one and the same act, the election of a member of Parliament, thus confusing political issues which should in the highest interest of the State be kept perfectly distinct. The attempted performance of these two requirements by the single process of election naturally results in the satisfactory attainment of neither. It causes general confusion, and has also the particular effect of introducing a disastrous instability into the administration of the country, by bringing about the fall of the Government in consequence of the rejection of some particular measure. This last most undesirable result has already been made the subject of severe criticism by leading publicists, though the more disastrous interference with the choice of the ablest administrators, entailed by the system in general, passes almost altogether unnoticed. The editor of the *Edinburgh Review* has pointed out that "the tenure of office of the executive government must be independent of the fate of particular legislative proposals": and he is eloquent on the evils arising from uncertain tenure of office, and the intolerable interference with the right of private judgment, which present procedure entails. With all due deference it may be suggested that the line of argument adopted by Mr Harold Cox has for its only logical conclusion not merely reasonable permanence of political tenure and liberty of political conscience, but government by the ablest. No reform in this matter can be considered as rationally complete which does not leave election free for its normal, natural, and original purpose—the recognition and promotion of political talent.

The conclusions of the present chapter may perhaps be summed up as follows:

(1) That election and representation are two distinct ideas.

(2) That the only admissible meaning of election, looking to the highest interests of the State, is election of the best.

(3) That representation has ceased to have any meaning but the representation of opinion.

(4) That the representation of opinion is the

means whereby democracy exercises control over government, and in some form or other is a necessary process.

(5) Under the present system, however, there is, at every general election, confusion between the idea of the election of the best and the desire to promote a certain theory of politics, or even to advocate a particular measure, and that this is detrimental to the highest interests of the State.

(6) That it is impossible under the party system to prevent the electorate from regarding the political opinions of a candidate as of more importance than his character and ability.

(7) That the destruction of the party system is therefore the ideal end at which to aim.

(8) That an advance in this direction may be made by preventing such a ludicrous travesty of the true principles of government as the election of a Parliament, not for the purpose of securing great political leaders and advisers, but for the purpose of deciding some question of the hour such as small holdings or Chinese labour, and by rendering the resignation of a ministry unnecessary on the loss of such particular measures.

(9) In furtherance of these objects it seems advisable that some device such as the referendum should be adopted, so that the registering of a democratic verdict will be no longer allowed to interfere with the true function of election—the recognition and promotion of political talent.

(10) Finally, an examination of the phenomenon

of election confirms the conclusion previously reached, that the evolution of government is the combined result of two great tendencies, one of which brings to the head of the State men of superior ability, while the other places them under efficient political control. The principle of election corresponds to the first of these processes, the principle of representation to the second. In election is implied the idea of discovering the statesmen fittest to be entrusted with great political duties, in representation that of sending up a body of men, duly qualified to watch the government in the interests of the people, for the purpose of securing universal political justice.

CHAPTER V

POLITICAL ASSEMBLIES

ON one of the rare occasions when Macaulay seems to distrust the all-sufficiency of his political discernment, he admits that "the limits which separate the power of checking those who govern from the power of governing are not easily to be defined." It is the object of the present chapter to show that those benefits which Parliament as a corporate body confers upon the nation must be clearly distinguished from the advantages which result from the triumphs of individual statesmanship. Certain functions which have hitherto been incorrectly attributed to Senates or Parliaments in reality belong to particular individuals, and while the importance of this individual influence has been overlooked, political assemblies enjoy a more widely extended fame and inspire a greater confidence than their actual performance warrants. From the political point of view there have been only two relatively successful nations in the history of the world—the Romans and the English. It was argued in the second chapter that their success was due to the fact that government by the ablest was in practice the characteristic political feature of both these peoples. Such a belief, however, in view of the importance usually attributed to political assemblies, involves an inquiry into the relation in which Senate or Parliament stands to the actual working of government by natural selection. This inquiry will reveal the fact already mentioned that many constitutional theories which meet with almost universal acceptance are nevertheless of the nature of mere speculative explanations or plausible assumptions, which are not found on examination to correspond with the real facts of history, or of political life as exhibited in history. It is true that many of these constitutional fictions have had a practical object, and in addition to other advantages which they have in this way conferred have no doubt been extremely useful in reconciling the spirit of conservation with the necessity of political change and progress. But they are fatal impediments in the way of the inquirer who is endeavouring to assign their actual value to the different influences which go to make up the total loosely described as government, and who in particular wishes to distinguish with reasonable precision between the forces which supply the driving-power of an administration on the one hand and the regulating power on the other. If we advance to the inquiry without any of the usual preconceptions, it would seem that a commonsense interpretation of the phenomena of assemblies will necessitate the abandonment of certain current beliefs as to the manner in which the various duties of government are apportioned and distributed. If we, in short, restrict ourselves to the definite and actual performances of Parliament, and concentrate our attention upon what does happen, rather than upon what ought to happen according to certain constitutional theories, the conclusions of the previous chapter will be confirmed. It will be found that both the existence and the functions of a political assembly are subsidiary to other forces which supply the initiative and real vitality of government. Parliament comes into existence not to conduct the government, but to control the action of those who do.

A political assembly, as already pointed out, is in its essential character a practical manifestation of the will of the people, and its function is to control the phenomenon of government in the interests of the whole nation. A high authority has said that the English Parliament in its origin was an "act of speech," the act of the king consulting with a few of his leading chiefs on some detail of policy. In such an act may be discerned as potentially present the whole theory of the "advice and control of Parliament." To permit advice to be given is by implication to recognize the right of the governed to modify the policy of the administration in their own interests. To this it may be added that throughout the early history of the English Parliament, the tone adopted is advisory and restrictive but almost never authoritative. The distinct impression which we derive from the actual facts of history is that of an assembly who recognize the active functions of Government as belonging to an authority which they claim to advise, to regulate, and finally to control, but which has an independent existence outside their own body.

Perhaps it will be urged that legislative power has always been attributed to Parliament and that this constitutes a proof of the possession of sovereign authority. It must be admitted that the function usually associated with political assemblies, which are indeed generally called legislative bodies, is that of making laws. Until a comparatively recent date the correctness of this view has not been seriously questioned. But as in the present inquiry nothing is taken for granted, it will be instructive to investigate the grounds upon which this interpretation of the function of political assemblies is based. The possession of legislative powers in any real and accurate sense by the two most famous political assemblies in history—the Roman Senate and the English Parliament—will be found upon examination to be subject to the gravest doubt.

It sometimes happens that a faulty generalization in the world of science or of politics hits the popular taste for easy definition, and is from that time accepted as a really scientific statement of the facts of the case. There is the strongest reason for believing that the term "legislative" is a question-begging epithet of this kind, at least as applied to the two great political assemblies which are universally supposed to have been pre-eminently legislative. The origin of this curious misapprehension seems to have been as follows:—A true division of the functions of government into executive and legislative powers has first been made. For it is obvious that what we may call the conduct of political operations is one of the duties of Government, while it is equally clear that the framing of laws for the regulation of the lives of citizens is another. But a hasty and inaccurate observation has wrongly assigned these different functions to different sets of individuals. The temptation to do so was no doubt very strong. In any considerable State there appears a body of men forming part of the Government, who are undoubtedly men of action. There is also another body of men who are not men of action, and who take no part in the executive duties with which other officers of the State are charged. What more natural than to suppose that this second body of men, whose activities are restricted to sitting in a large room for the purposes of oratory and deliberation, and who in addition have always been called a legislative assembly, are the intellectual authors of the legislation of the country, while the men of action, the executive, carry these laws and decisions into effect. The functions of ministers and the functions of councils are in truth different. But that this difference arises from a division between them of executive and legislative duties is a view unsupported by the real facts. So far as the present author is aware, Sir John Seeley was the first who emphatically declared, as the result of life-long observation, that the legislative power in any real sense belonged far more truly to the Cabinet than to Parliament. It is the Cabinet who decides what measures are necessary, it is they who put them into shape, it is they, in fact, who initiate legislation, determine its character, and place it before the Parliament in a form provisionally complete. Parliament amends this legislation or rejects it, and the whole of their so-called legislative duties are summed up in these alternative operations. It follows that those who arrange the measures which are to be introduced, and the form that legislation is to take, without submitting this preliminary procedure, which is far the most important, to Parliament, are the real authors of the legislation of the country. The law, in fact, is made before it reaches Parliament. Parliament decides whether or not it shall be placed on the statute book, but this is the power of veto, not the faculty of legislation. Mr A. L. Godkin, in referring to the functions of the English Parliament, comes to the same conclusion as Sir John Seeley. Indeed, the fact would seem to be sufficiently obvious, when once the mind is free from any arbitrary preconceptions as to the precise portions of political anatomy in which the legislative and executive functions reside. Of the English Parliament he says: "Its proceedings are controlled and regulated by the executive officers. They submit measures to it, and ask its advice and consent. . . . The function of the two houses is essentially not the draughting or proposing of laws, but seeing that no law is passed which is not expedient and constitutional. . . . The British ministry in fact legislates as well as administers"; and again, "The function of the British ministry is to provide the necessary legislation."[1]

Precisely the same criticism may be passed on that other great historic assembly, intimately associated in the popular fancy with legislative functions, the Roman Senate. It is called a legislative assembly, just as the English Parliament is called a legislative assembly, and just as most political assemblies are called legislative assemblies, in conformity with a preconception which will not bear examination. "Its chief distinction was that, as a rule, subjects for consideration on which it had to take positive action did not originate with it, but were brought before it by the executive officers engaged in the active conduct of government. So that it may be called a consultative rather than a legislative body."[1] Nor was there any greater legislative power in either of the two assemblies through which the Roman people exercised their sovereign prerogatives, the *comitia centuriata* and *comitia tributa*, since neither could act unless set in motion by a magistrate.

[1] "Unforeseen Tendencies of Democracy," pp. 99, 108.

We must now see what is the bearing of these conclusions upon the theory of the present work. It would seem that the reputation which assemblies have gained as the true repository of the most important powers of government is by no means justified with regard to those legislative functions with which their activities are especially associated in the popular mind. If neither the Roman Senate nor the English Parliament are found upon examination to be really legislative bodies, if the two most famous assemblies in the world, whose existence is habitually identified with the making of laws, are found in reality to exercise no such function, then the plausibility of the view that political assemblies in general supply the motive force of the active powers of government is very seriously diminished. It follows, therefore, that the value and impressiveness of that department of government to which the faculty of legislation can be traced is correspondingly increased. It would seem, in fact, that not only the administrative but the legislative faculties are to be found in the hands of those few individuals whom, according to the present theory, the course of natural selection has placed at the head of the State. The whole, or nearly the whole, of the power and dignity which has been attributed to Parliament, in virtue of the belief that it was the source of legislation, belongs in reality, as the theory of this work would have led us to anticipate, to the initiative of the individual rather than to the corporate action of a heterogeneous political assembly. The influence of the administrative few is seen to be even more far-reaching than was at first supposed. Administration includes legislation. The legislative as well as the executive powers of the State are found to reside, not in the collective wisdom of a Parliament, but in the mind of dominant individuals.

The view that real legislative power does not belong to Parliament is confirmed in another way. The belief that a "constitution" is a sort of political labour-saving machine, which will give the same beneficent results in any country where it may happen to be set up, has led to the attempt to reproduce the British constitution in many countries on the Continent, in the confident expectation that it would confer the same advantages so long enjoyed by the English people. And this it conceivably might have done had the essential features been accurately or even approximately copied. The authors of the continental imitations, however, proceeded on the assumption that the secret of the vitality and comparative excellence of English administration lay in a certain definite arrangement and division of parliamentary functions. They missed the essential truth that the true cause of governmental success is to be found, not in the adoption of a certain kind of parliamentary assembly, not in a cunningly devised system of checks and balances, but in the facilities given, over long periods of history, for the predominance of individual talent. Furthermore, believing that the art of government consists in the art of legislation, and that assemblies are the real authors of legislation, which they frame and pass after due discussion, those countries who wished to follow in the political steps of England imitated what they imagined to be the actual process at work in the mother of Parliaments. The failure to produce anything like the same results is a strong proof that some vital element has been missed. The theory that assemblies are the real source of legislation does not work out in practice. "When the House of Commons was taken as a model on the continent, what was set up was not really the English parliament, but a set of councils for discussion, in which every man had the right of initiative . . . as a rule the continental legislatures, while modelled on the British or cabinet system, have really not copied its most important feature, the dominance of the executive in the legislative body."[1] The general result of these attempts to identify Parliaments with legislation has been endless talk and incessant changes of administration, due to the efforts of each amateur legislator to push his own private fads. But the vitality of government, the coherence of legislative aims, and the general continuity of policy, which is necessary for national success and which depends upon the legislative and executive predominance of a single set of individuals, has been entirely wanting.

[1] Godkin's "Democracy," pp. 102, 103.

The same unsatisfactory results have been brought about in America, but by a curiously different train of circumstances. That separation between legislature and executive which is the cause of the trouble, and which has taken place in continental countries accidentally, so to speak, through their failure to reproduce the essential features of the English parliamentary system, has been produced in America with deliberate intention The authors of the American constitution seem to have been fully aware of the power of the executive over what they thought was the legislative department of the English government. Unfortunately for the future of the American constitution and people, they came to the conclusion that this power was contrary to the interests of the community, and amounted, in fact, to a baneful perversion of the arrangement proper to an ideal constitution. For this conclusion they had, to all appearance, good and satisfactory reasons. They were impressed by the undoubted mischief and confusion which the arbitrary conduct of George III. and his ministers introduced into English politics at the beginning of his reign, and they determined if possible to avoid the recurrence of similar troubles in America. With this end in view they determined on the separation between the legislative and executive powers, or to put it more accurately, they determined on the separation between the executive and the American assemblies which reproduced our Houses of Parliament. They accordingly enacted that "No person holding any office under the United States shall be a member of either house during his continuance in office." As Bryce says, "This restriction was intended to prevent the President not merely from winning over individual members of Congress by the allurements of office, but also from making his ministers agents in corrupting or unduly influencing the representatives of the people, as George III. and his ministers corrupted the English Parliament."[1] Or again, as Mr J. A. R. Marriott says, they preferred to follow the theory of Montesquieu rather than the practice of Walpole. They can in truth hardly be blamed for failing to perceive that the faults of which they complained, and which they wished to avoid, were the accidental results of a momentary situation, rather than inherent evils of the system which closely associates those prominent individuals who "administer" with the larger body who "deliberate." Still less could they be expected to anticipate a truth almost universally unrecognized at the present day, that "the dominance of the executive in the legislative body," so far from having a prejudicial influence on the interests of the State, is an essential condition of healthy government. The weakness and confusion which this separation has introduced into American administration will be pointed out in the succeeding chapter.

The broad results of this inquiry into the

[1] Bryce's "American Commonwealth," vol. i. p. 87.

supposed legislative powers of assemblies may be summed up in the following way :—There is nothing to prevent an assembly from occasionally constituting itself the legislative power of a nation; there is no physical impossibility which incapacitates a council from framing legal resolutions and making them the law of the land. It could no doubt be proved from history that assemblies have actually, and not merely in theory, done so. Still less is there anything to prevent a Parliament from firmly believing that it is the true source of legislation, and finding lawyers and constitutional historians to support this view. When, however, we approach the question from the point of view of the practical working of a constitution, we find in the first place that the efficiency is greatest where, as in the case of the Roman Senate and the English Parliament, the real legislative capacity is in the hands of the executive. In the second place we find that where this truth has been deliberately controverted, as in the case of the American constitution, or accidentally ignored, as in the case of continental legislatures, the consequence has been legislative failure and governmental impotence. The conclusion, therefore, which has already been reached by a different line of argument, is thus confirmed, that in the evolution of government and of a political constitution, men take the precedence of measures. A country profits more by the leadership of talented statesmen than by the misplaced legislative activity of even the most conscientious political assembly. Adjust political selection so as to secure men of the best available quality and the measures will tend to be of the best available quality also. To the supervision, possibly to the improvement of that quality, Parliament will contribute, but will not do more. The verdict of nature and of history in the sphere of legislation might be summed up in the words, "Take care of the men and the measures will take care of themselves." To concentrate attention on measures without much regard for the men who guide the State is a procedure as unreasonable as to provide the most perfect and the most complicated machinery on board a battleship, and to ignore the capabilities of the men behind the guns. When we come to study the matter without any preconceived determination to follow out the lines of demarcation laid down by Montesquieu and others, if we ignore this almost *a priori* assumption that an assembly which is called legislative must necessarily be the source of legislation, it seems entirely reasonable that the power of legislation and of administration should be combined in a single mind. Any attempt at distinction between the two faculties is academic and unreal, for they are merely different sides of one political genius.

Finally we come to the elective functions of political assemblies, important both in the case of the Roman Senate and the English Parliament. It will be shown that it is from their elective powers that these assemblies have gained much of their renown, not because of what election does, but because of what it seems to do, a conclusion for which the previous chapter has already prepared us. The simple process of crediting these assemblies with the glory of the statesmen and ministers whom they elect, has in fact given them a reputation for administrative ability to which in their corporate capacity they can lay no claim whatever.

The power of making a good choice is a valuable faculty, but however valuable, it cannot create the qualities which it merely recognizes. A king, let us say, is elected by the Witan, and leads his followers to victory. Such a transaction, however, is frequently spoken of as if the choice of the Witan and not the strategy and tactics of the king were the actual cause of success. But it is clear that an assembly does not create in a king or prime minister any of the qualifications which it utilizes in the interests of the community. The most venerable and celebrated Parliaments in the world have never done more in their elective capacity than make use of the faculties of distinguished individuals. The will of the assembly has nothing to do with the making of the qualities of the man. Their greatest power or their utmost good fortune is limited to a capacity for recognizing a great man when they see one. Stripped of all unnecessary mystification their elective power resolves itself into the commonplace process of putting the best men in the most important posts. The decisive factor is the self-determined ability of the leader, not the merely nominating faculty of the assembly. Yet some historians manage to convey the impression that Parliaments not only appoint a minister but are in some mysterious way the authors and creators of his merit; and half the reputation of political assemblies in history has been built up by the process of deliberately usurping credit for the performances of gifted individuals. Suppose the question to be the election of a general to conduct a decisive campaign, or of a minister able to appease factious discord and secure the unity of the nation in some vital political crisis. The will of the assembly cannot create a great military leader with invaluable strategic faculties, nor can it endow a statesman with the genius or powers of administration necessary to save the State from dissolution. From this point of view an assembly is an utterly barren and unproductive political phenomenon. Having regard to the actual accomplishment of the work, the general who wins the campaign or the statesman who saves the State is surely a more important factor in the result than the assembly who requests the general to be so good as to win the campaign, or who graciously permits the statesman to use his best endeavours to save the State, Yet even Mommsen gives the Roman Senate credit for great administrative acts, which they merely commissioned others to perform. Historians, it is true, recognize when necessary the existence of an executive as a separate feature in the government of a nation. But this does not prevent them, by a species of loose political thinking and writing, from incessantly attributing or seeming to attribute, to the Roman Senate and the English Parliament the individual deeds which built up Roman or English greatness. Among other ambiguous sentences of Mommsen which countenance this view we find the assertion that the Senate was called upon "to rule great bodies of subject populations and carry on foreign wars." Pelham, however, is much nearer the truth when he insists that "nothing is more characteristic of the Roman constitution than the width and completeness of the 'power of command' with which the Roman people invested their chief magistrates. The magistrate with the imperium was in theory and for the time all powerful."[1] Granting that this freedom of individual initiative was not altogether maintained, and was indeed at one time seriously impaired by the encroachments of senatorial authority, yet the corporate action of the Senate never actually replaced, and never could replace, the individual action of consuls, proconsuls, and prætors. We have only to come to close quarters with this notion, that an unwieldy body such as a Senate or a Parliament governs an empire, in order at once to perceive its inherent absurdity. No Senate was ever known to command an army or even to conduct a campaign, no assembly was ever known to administer a distant province. It is

[1] Pelham's "Roman History," p. 151.

absurd to speak of assemblies governing and ruling when their activities are restricted merely to supervising these operations.

The reputation of the two greatest assemblies in history has resulted, not from their character as legislative, still less as administrative bodies, but simply from the fact that they included in their numbers those active and dominant individuals who are called the executive of the State. They shine with the borrowed light of these great men, men whose functions are not the controlling and amending functions which characterize the assembly in general, but the functions of leadership and initiative, which differentiate them entirely from the other members among whom they happen to sit. The true explanation of the greatness of Rome and England, in so far as it is identified with senatorial and parliamentary government, lies in the fact that these assemblies gave free play to the individual genius of the distinguished men among them. The great triumph of English constitutional history is generally supposed to be the establishment of the responsibility of ministers to Parliament. It is not usually remarked that in securing this necessary and valuable stipulation the way has been opened for the operation of the reverse principle, the influence of the executive upon the assembly. Whatever the motives may have been for insisting upon the presence of ministers, according to the present view the most important result, both in Roman and English history, was not the subordination of the ministers to the assembly but of the assembly to the ministers. The conclusions of Mr A. L. Godkin on this subject are so important, formed as they have been from an entirely different point of view, that no apology is necessary for quoting once more from his well-known work. "Throughout the whole course of Roman history the administrative officers remained actually in charge of the government, subject to the advice and control of the legislature."[1] "The same system has prevailed in the British Parliament ever since it became a real power in the State. Its proceedings are controlled and regulated by the executive officers. . . . The House of Commons never reached the point of seeking to take the administration of the government out of (the king's) hands or out of those of his officers except in the rebellion of 1640. Its highest ambition was to be consulted about what was going to be done, and to be allowed to ask questions about it, and to vote the money for it. It never thought of taking on itself the function of administration. It confined itself to the exercise of a veto. The ministry never parted with its power of initiation . . . the English House of Commons, one may say, has grown up under the consultative system. No other system has ever been seen or thought of." Such is the record, we might almost say the self-effacing record, of the two greatest political assemblies in history. No nation can

[1] Godkin's "Democracy," pp. 98, 100 and 101.

hope for continued success unless it restricts its national council in a similar way, leaving difficult decisions in matters both of internal and external policy to those individuals who, under healthy political conditions, will always rise to a well-deserved national supremacy. It is by the exercise of this infinitely wise moderation that the mother of parliaments has conferred greatness on the country and the Empire.

The question now arises: Does Parliament fulfil any other useful duty besides that of directing government in the interests of the nation as a whole? The impression which the existence of political assemblies has made upon learned and unlearned alike seems to point to the conclusion that they perform some other important function of which the value is generally felt, though its exact nature has escaped observation. Accordingly the suggestion is here put forward that political assemblies have a meaning and importance which is connected with the efficiency of government in a way not usually recognized.

The individuals who compose a national assembly appear under two different aspects, and indeed present two different characteristics, according as we consider them in relation to their collective duties or in relation to one another. The Roman Senate or the English Parliament may be regarded either as a single body engaged in certain political functions or as a series of individuals before whom the prospects and prizes of political life are competitively displayed. As a collective body Parliament acts on behalf of the nation, while in the other case it is resolved into a number of rival politicians, keenly engaged in the struggle for political distinction. Under these latter conditions a double process may be observed in operation which has a two-fold influence upon the position of those who take a prominent part in the doings of the assembly. As a result of individual rivalry a strenuous political selection takes place, while at the same time a certain amount of political training is imparted to those members who are destined to become the statesmen of the future. On the theory that national success is largely dependant upon the intellectual quality of the administration, the competition of a picked body of able and energetic men, all anxious for celebrity, is an excellent method of ensuring a high intellectual standard of governing ability. The meaning of this phase of the activities of a political assembly will perhaps be more readily recognized when the reader is reminded that, as shown in the preceding chapter, such competition is only another aspect of the familiar phenomenon of election, one of the most important functions of the House of Commons. The process known as the formation of a Cabinet is, when stripped of constitutional embellishments, the outcome of this unacknowledged rivalry between men of commanding character and intellect who have all enjoyed the advantage of long political experience; it is the triumph of a trained statesman formally recognized and approved by clear-sighted colleagues ; and this is the reason why it can show better results than are attained under the American system, where the selection of a President is made by a populace who are comparatively ignorant, and who in addition consider the exercise of their sovereign powers of more importance than the capabilities of the statesman upon whom their choice may happen to be fixed. Cabinet election is the culminating process in that struggle for political precedence which has never yet been suspended in any so-called legislative chamber in the world. The silent contest between prominent men on the same side of the House is in reality more keen than the noisy warfare between the leaders of the opposite parties, because in the first case the triumph of a rival means the loss of a career, while in the second case it is merely a victory for the common enemy, which leaves the question of personal supremacy unaffected. It is difficult to harmonize the popular view of the functions of the House of Commons with this and other symptoms of an activity totally irrelevant for the performance of those duties which are supposed to supply the reason for its existence.

According to the popular theory, an assembly is composed of grave and reverend senators, consulting earnestly together for the high interests of the State, and continually occupied in devising and passing laws which have for their sole object the promotion of the general welfare. When, however, we come down to facts, it will be admitted that, wherever a philosophic atmosphere of calm political reasoning may exist, it does not exist in the House of Commons; and that keen personal rivalry as well as incessant party faction is a conspicuous feature of a chamber which should in theory be a quiet retreat for the discussion of measures for the furtherance of social and political well-being. This vivid intrusion of personal and party ambition, so irreconcilable with the theoretically accepted view of the nature of a political assembly, has a new significance in the light of the interpretation suggested above, especially when we see that the actual result is to subject selected political talent to a further process of selection.

But perhaps it will be urged that in proportion as this point is made good, the idea of Parliament as a disciplinary institution for the training of statesmen is rendered ridiculous. Those who recall the tempestuous conduct and bitter language of many historical debates may be inclined to smile at the association of such scenes with anything in the nature of education in political principles. They would, however, be wrong to underrate the educational value even of apparently fruitless and acrimonious parliamentary wrangling. From the present point of view one of the greatest advantages of English public life is the fact that keen political rivalry can exist without necessarily involving personal and individual animosity. This is what the parliamentary system with all its grievous shortcomings still manages to effect. It may even be urged that this counts for more than a scientific training in the principles of ideal government. What a writer in the *Quarterly Review* calls "an inherited aptitude for the compromises necessary to common political action" is in reality the result of centuries of such parliamentary discipline. In those countries which have endeavoured to assimilate the principles of the British constitution, the irreconcilable nature of the jealousies of rival politicians has fatally hampered the efficient working of government. It is useless to grant free institutions, as they are called, to countries that are politically untrained, and then to expect that constitutional devices will transform, as by magic, the inherent deficiencies which arise from such want of training. The comparatively smooth working of the system of individual rivalry in English politics cannot be dissociated from the training received by English politicians during a long period of strenuous parliamentary discipline.

If, then, any additional reason beyond the value of its controlling powers is necessary to explain why the existence of a vigorous political assembly has always been an indispensable accompaniment of national success, that reason is probably to be found in the fact that such an assembly affords a valuable supply of trained political talent upon which the Government is able to draw at will. This view, however, of Parliament as a medium of political selection, though of much interest to the general theory of this work, is no necessary part of the present argument. The essential fact is that Parliament exercises a moral and financial control over government but does not itself govern. The real governing power of the nation is to be found in the administrative talent which by a natural process of selection is placed at the service of the State. Towards this individual talent Parliament occupies the same position that it formerly occupied towards the king, permitting the enjoyment of distinctive powers, but on the sole condition and with the sole object of promoting the national welfare. The true division of the functions of government is not into legislative and executive, but into a department which conducts the government, legislation included, and a department which exercises a general supervision, and punishes any misuse of the powers which are entrusted to leading statesmen. In support of this conclusion, which has been formed by a dispassionate study of the actual work of statesmen on the one hand, and of political assemblies on the other, the attention of the reader may once more be drawn to the result of the investigation in the preceding chapter. It was there pointed out that the two different phenomena of election and of representation suggest a division of the functions of government such as is here maintained, between certain individuals who act and certain others who impress upon them the general wishes of the community. In the present chapter it has been found by a perfectly independent analysis, free from the distorting influence of the supposed necessities of constitutional theory, that the division of the duties of government between statesmen who conduct and an assembly which controls, is a natural inference from the facts of history. The difference between election and representation corresponds to the difference which has now been shown to exist between the active work of statesmen and the restrictive influence of the political assembly. Why these two ideas are not in practice kept perfectly distinct has already been explained. Under ideal conditions a Parliament would consciously select its ablest men to grapple with the intricate problems of domestic and imperial statecraft, while the remaining members would recognize that their functions as a body were of a distinct and subsidiary character, mainly designed to ensure that the benefits of government should be equally distributed to all. And this is what actually takes place under the English political system, though imperfectly and in an unconscious and informal manner. Of the members sent to Parliament by a similar process of election, some eventually exercise leadership, while others remain associated with the idea of supervision, which sums up in a word the real duties of a political assembly. It is the function of Parliament as a corporate body to watch over the happiness of the people as a whole, to oppose any attempt to govern in the interests of a single class, to visit with extreme displeasure the want of reasonable political foresight, to criticize and control, but not to conduct, the operations of government.

CHAPTER VI

DEMOCRACY

THOSE who are imbued with the idea that the predestined end of all political progress is to place the governing power entirely in the hands of the people, will regard the present theory as irreconcilable with their cherished republican ideals. That the tendency of political evolution is towards a certain kind of democracy need not be denied: the only question is what form that democracy will assume. It will be shown in the present chapter that whatever else the will of the people may achieve, it does not suspend the rivalry of individuals for political distinction, nor does it abolish the intimate connection between national welfare and government by the ablest. Popular government has added strength and glory to a State only so long as constitutional innovations have not interfered with the natural supremacy of talent; and if republics have ever shone with political brilliance, it has in reality been from the reflected light of distinguished statesmen who have been permitted to hold office. The rule of the many has only been successful when by abolishing privilege they have assisted the rise of able men to political power, and have abandoned for the time any attempt at "government by the people." The only triumphs of democracy are those which it has gained by facilitating the course and adopting the principles of government by natural selection. Conversely, where the rule of the people has led to ruin, it has been in consequence of a flagrant disregard of the conditions described in the previous pages as essential to national success. Republican government comes into existence chiefly for the purpose of remedying the deficiencies of monarchy. Yet it fails in consequence of a repetition of the cardinal errors of the system which it summarily rejects. Inspired for the most part by greater earnestness of purpose, republics can show no striking superiority of result, because they have not wholly understood the value of the political disorder for which they seek a cure. The evil of absolute power has not been traced by the republican physicians, in one respect at least, to its real and fundamental cause. The treatment has been directed to prevent the dissipation of the resources of the State by royal extravagance, to abolish the royal claim to be above the law, and to forbid arbitrary interference with the liberty of the individual. Yet when all these excellent stipulations have been enforced, there still remains a further radical improvement necessary for the continued welfare of the State. The essential mischief caused by absolute monarchy, as we have so often said and must once more repeat, does not so much result from any interference with the general liberties of the people, as from the attempted performance of duties which are beyond the generally limited capacity of hereditary rulers of a State. This same identical vice reappears, with certain variations and in a slightly different form, but with equally fatal consequences, under democratic government. Like the monarchies whose memory they affect to despise, democracies have thwarted the access of ability to power, and have almost consistently hindered the employment of conspicuous talent in the highest offices of the State. Rarely for any length of time have they been found to accord willing recognition or ready encouragement to those of their fellow-citizens who have given evidence of the possession of an offensive superiority of political genius. On the contrary they have, after the manner of Eastern potentates, preferred to read the evidence of their own selfish power in the dependent nature of the instruments of their policy. In the light of the present theory this well-known failing of democracy acquires an entirely new significance.

In order to explain how institutions so different as democratic government and absolute monarchy—the one professing to aim at the greatest happiness of the greatest number, and the other asserting a right to arbitrary power—have both agreed in the commission of the same offence, we must refer the reader once more to the conclusions which were reached in the chapter on election. It was there found that election is the recognition of superiority. Election introduces into the evolutional process in politics no change of principle but merely a change of method. It consciously performs that great task of political selection which before took place unconsciously as the result of ambition, or interest, or patriotism, or all combined. By this change of method a very great difficulty has now apparently been overcome. Previously political progress was greatly hindered by the fact that the right man did not always find his right place, and the supreme man did not always win supremacy, because the system of natural selection in politics is neither free from disturbing causes nor perfect in the achievement of its results. If, however, election by the people becomes the recognized method of authorizing the exercise of political power, an intelligent desire to ensure government by the fittest might then be expected to take the place of what was formerly but a tendency of nature, liable to be hindered by many causes and especially by rulers unconscious of its usefulness. It would seem, therefore, that the democratic preference for election is the consummation of the evolutional process, and that in this final stage of political development no serious obstacles will prevent the leadership of the wisest and the best.

Unfortunately, though efficiency is the only justification of election to administrative office, the adoption of the purely elective principle has been found not more but less conducive to really able government. The disturbing causes which thus divert election from its true purpose are many. Electors are human and very liable to misjudge political capabilities. They frequently prefer the safety of commonplace attainments to the dangers of political originality; they blindly worship eloquence; they are attracted by those they admire rather than by those whom the situation demands; they are captivated by qualities which add nothing to national strength; while jealousy of real talent may actually impel them to elect those of inferior capabilities. But strangely enough the defect which most seriously impairs elective efficiency is one which is allied to the highest republican qualities. The belief to which reference has already been made, that all misgovernment is the result of injustice, and that so long as rulers are well intentioned, their mental capability is an entirely subordinate consideration, has especial consequences for a democracy, since a republic is nothing else than an attempt to carry this theory into actual political practice. It follows from this assumption that if the work of government is performed by men determined to carry out their duties, not in a spirit of self-aggrandisement, but under a solemn sense of obligation to the community, all those evils which are the result of tyranny, love of power, and selfish ambition, will as a natural consequence disappear, and the State will be enabled to enter on that steady course of peaceful progress which has only been prevented by the dishonesty of the governing powers. An expectation founded on so complete an ignorance of human nature was by itself certain to experience the most complete disillusionment. But when we find this confiding reliance upon mere rectitude of purpose combined with the most determined adherence to another republican dogma, the right of perfect freedom of election, we arrive at a combination of fallacies bound, when translated into practice, to produce the most disastrous political consequences. The conviction, perhaps we may even say the honest consciousness, of a superior excellence of intention, which places the welfare of the community before all personal considerations, seems to produce a kind of political conceit, closely resembling self-righteousness and unctuous godliness in religion, which assures the democrat of the wisdom of all his acts, and which especially seems to enforce the conviction that a Parliament of right-minded patriots can make no mistake in the selection of the officers of State. Aware that some degree of competence is necessary even in the ministers of a republican assembly, yet at the same time determined never to admit that the exigencies of the situation can impose any limits to the freedom of their choice, they triumphantly solve the difficulty by assuming that any person whom they may select is capable of administration simply because they select him. It is not so much that they ignore efficiency, or reject it for a different standard, as that they beg the question of efficiency in favour of any instrument chosen by themselves: *vox populi, vox Dei,* in its latest form. We have said that election is the recognition of superiority. Democracy, to use a current phrase, goes one better, and declares that election creates superiority. The true democrat, like the ardent lover, endows the object of his affection with virtues and capacities which are in the highest degree improbable. With the duties which democracies impose, they imagine they confer, by the mere fiat of their will, the power to perform them. When a supremacy is registered in the course of natural selection, reality of merit goes hand in hand with elevation to honour. The elective principle, when exercised by an assembly, must likewise keep in close touch with the requirements of national life and the exigencies of the international situation, if it is to justify its place in the evolution of politics. Unfortunately, however, the exercise of perfect freedom in election may confer a supremacy which has indeed the authority of the multitude behind it, and which is calculated in consequence to impress the unthinking world, but which has no real justification in the character of the recipient. We have, in fact, under these circumstances to distinguish between two kinds of supremacy—one which is justified by the required personal qualities and one which is not, and which therefore may be called merely "elective supremacy." The power of the people, conferred upon an unsuitable recipient, constitutes an elective supremacy. The distinction is one of the greatest importance to the present theory, because it is this "elective supremacy" which, by usurping the functions of a natural supremacy, causes the ruin of democracies.

A remarkable instance of the belief that the mandate of the people invests the privileged individual with magical powers of performance, and that election can create the necessary qualifications in defiance of the actual requirements or reasonable probabilities of the case, is to be found in the action of the French revolutionary assembly. This delusion gave rise among other calamitous consequences to a mistake which seriously hampered the French Republic in its struggle with England. One of the first results of the Revolution was the retirement or dismissal of the aristocratic Royalist officers of the navy, the majority of whom were endowed, not merely with high professional ability, but with the invaluable experience gained in the recent war with England. The navy was thus reduced at one blow to a state of disorganization and inefficiency due to the professional ignorance of the new revolutionary officers, since fervour of republican sentiment, instead of expert knowledge of one of the most difficult of professions, was made the qualification for naval command. The English navy had in consequence a comparatively easy task where the severest contest might have been anticipated, and but for the invigorating moral effects of this initial triumph it is quite possible that the nation might have faltered beneath the terrible burdens which they were subsequently called upon to endure. This naval success was the literal gift of deluded democrats, who in their fatuous self-conceit imagined that the mere will of the Republic, the mere order of a revolutionary committee, could replace the almost priceless weapon they had deliberately thrown away.

In the history of Holland and the Netherlands we find an additional error, well calculated to exaggerate the evil consequences of those elective vagaries which arise from a too robust consciousness of political infallibility and rectitude. One of the prevailing sentiments in a typical democracy is the fear of individual greatness. It may perhaps be true that this feeling is not to be ascribed, as some allege, to the meaner instincts of human nature, but to experience of the fact that in the history of government great endowments of mind and character have frequently been found associated with a tendency to abuse power. This circumstance, however, does not alter the unfortunate results entailed by a distrust of ability in politics, since one of the first uses to which democracy puts the right of free choice is to select not statesmen but delegates whose submissive attitude and limited mental powers render the recurrence of tyranny a remote political contingency. Intellectual mediocrity thus becomes an actual qualification for office, in complete disregard of the truth that talent remains indispensable for political and military success. Freedom of election in particular enables politicians to indulge to the full their civilian apprehensions of military greatness, and thus dangerously to weaken the international position of the community. These two principles, the belief in the right of perfect freedom of election, combined with the fear of individual and especially of military greatness, may be assigned as the chief causes of the ruin of the Netherlands. Though there were other errors which do not directly concern the present argument, the selection of delegates with qualities which promised democratic safety rather than real political competence was the most fatal error of all. Even where the rise of military or naval genius imperatively required by a dangerous international situation could not be altogether prevented, the democratic authorities, apparently in order to demonstrate the fact that no unworthy considerations of military efficiency should be allowed to interfere with the freedom of civilian choice, took care to neutralize the talents of the commander by the appointment of a civil superior.

The same disastrous fallacy of the virtue of free choice, implying the belief or rather the obsession that the people can by election call into existence the talent necessary for the adequate performance of the most vitally important duties, was also the chief cause of the decay of Holland in the seventeenth century. This fundamental defect, to which all their misfortunes may finally be traced, is not without interest for the English people at the present day. Our own politicians also claim the power to create military or naval talent at will, and the vote of a party majority is in our own case warranted to convert the most unlikely characters into the supremely competent organizers of the armed power of a mighty empire. Yet what could be more ridiculous than to suppose that the constitutional necessity of subordinating military to political requirements, which confers upon politicians superior constitutional powers over the naval and military departments, also invests them at the same time with supreme strategical knowledge! An even greater disregard of common sense was exhibited in the wars of the Dutch against the English, and the general want of administrative efficiency which follows when election is not directed to its proper and rational object is fatally evident. The self-satisfaction of the Dutch democracy was such that their chosen political leaders felt themselves fully qualified by virtue of the democratic mandate to conduct the military affairs of the republic, with the result that the incompetent administration of a civilian government [1] "utterly neglected the navy, denuded the magazines, and was unable to produce at the critical moment the material necessary for battle." In the case of Holland and the Netherlands, as in the case of France, belief in the supreme merits

[1] Barker's "Netherlands," p. 282.

of free democratic election, almost as an end in itself, certainly as a process amply sufficient to secure political and national well-being, without the aid of the proper qualifications, had the effect of elevating to the most important positions men who were incapable of performing the supremely difficult duties which they lightly undertook. Election, rationally considered, entails the necessity of choosing the fittest for the particular purpose in question; perfect freedom of democratic choice, on the other hand, disregarding the question of fitness for the reasons already given, leads under existing international conditions straight to the ruin of the State. The fatal error here described as the institution of "an elective supremacy" is nothing else than the logical offspring of the supposed axiom that freedom to choose whom they wish is the most essential prerogative of democracies, and the sure secret of political success. It was in consequence of the determination to live up to the spirit of this strange metaphysical conception, while at the same time ignoring considerations of national efficiency, that the Dutch ruined their country. In many respects, if not in all, they were earnest and noble-minded republicans. Republican earnestness, however, is of little use for correcting a fatal perversity in the choice of national leaders. Liberty seems of but little national advantage when it issues, as it did with the Dutch, in a misuse of elective power which the most absolute monarch could not exceed.

What chance had a navy so officered, and a nation so governed, against Cromwell and his subordinates, one and all holding their respective posts in accordance with the strictest interpretation of the law of natural selection in war, administration, and government!

We have seen in the case of the French and Dutch republics the fatal consequences of the assumption that he who is supreme by the vote of the people is supreme in the world of fact. A still more famous instance of this delusion is to be found in the case of the Athenians. The history of the Athenian republic is indeed a signal illustration of the truth that democracies are successful only so long as the conditions favour the rise of great men to political power. It was the modesty of Pericles which prevented him from perceiving that the reliable working of the Athenian constitution was absolutely dependent upon a personal supremacy such as his own. The championship of democracy by Pericles is indeed one of the most formidable arguments in its favour ever brought forward in the history of the world, and those who refuse to believe that the great Athenian statesman was altogether mistaken, have some justification for their faith, but not that which is usually supposed. In all probability Pericles had divined the truth that certain conditions of freedom favour the rise of political greatness, and this it was which gave him confidence in the future of the Athenian state. He had, in fact, discerned the value of the free career in politics as tending, especially in Athenian life, to lead to the beneficial supremacy of talent and genius. On this hypothesis it may be assumed that he meant by the general direction which he gave to the constitution, to encourage an equality of political opportunity, in the perfect confidence that the adoption of such a course would ensure a constant succession of great men, capable of permanently securing the place of Athens among the communities of the ancient world. If his speeches in Thucydides are to be considered as a fairly accurate representation of his policy, they contain suggestions that this point of view is correct. The fellow-countryman of Solon, Cleisthenes, Miltiades, Themistocles, and many other distinguished men might well have hoped that the tradition of genius would be unbroken, and that the people could be safely trusted to facilitate its access to political power. The career of Alcibiades, possibly the superior of Pericles, in sheer intellectual ability is proof that the talent was there, and that it needed only a steady and self-restrained electorate to put it to its proper uses. But this second condition was exactly what did not exist. If any such scheme as is here suggested were really in the mind of Pericles, he failed to take account of the fatal tendency of democracy to defeat the natural law which brings great men to the front, by conferring an undeserved supremacy upon popular but worthless candidates. It was imperative for the success of such a scheme that the Athenian people should play their proper part of intermediaries rather than of principles, should assist rather than defeat the law of political selection, and should exercise the elective principle, not to gratify a sense of power and freedom, but to assist the upward course of genius to the highest offices of the State. In the sequel, as we know, they proved unequal to the task, and by the prejudiced selection of unworthy leaders, by the rejection of cautious and matured counsels, by indulgence in the light pursuit of varying democratic fancies, and the hasty execution of ill-considered political schemes, they involved the most famous of free constitutions in the political ruin which only a continuation of wise leadership can permanently avert.

Yet though the Athenian Republic ended in a rather pitiful exhibition of political frivolity, it is not its light-mindedness which constitutes the real danger of democracy but its earnestness, combined with a mistaken view of the nature of its own functions. Though absolute monarchy and democracy may differ widely in other respects, they are identical in this, that an overweening sense of their own importance has led them both to assume, on entirely insufficient grounds, that they possess not merely the right but also the ability to govern. Without the necessary qualifications for their self-imposed task, they usurp the place of really able men; or if they are at any time compelled to employ individual talent, they regard it not as the main source of administrative efficiency, to which their own functions are, or should be, entirely subordinate, but as a mere detail in the work of government, deriving its power and influence entirely from themselves.

These instances of the failure of democracy, however, belong to the past. The more enlightened democracy of the present day may perhaps be expected to approach their task with a wisdom which has been deepened by historical experience. A short review of modern democratic hopes and pretensions in the light of the present theory will enable us to ascertain how far these anticipations are justified, and in what respect and to what extent the principle of democracy is the principle of the future.

There is a general consensus of opinion that the tendency of modern times has been by a slow but irresistible movement to place power in the hands of the people. Both those who applaud this tendency and those who regard it with suspicion are at all events agreed upon the fact. A great change of feeling, the result to all appearances of some deep-seated social instinct, has brought about the admission of the lower orders to a political consideration which they never before possessed. It is, however, quite possible to admit that the so-called democratic movement has a deep evolutional significance, and yet to affirm that current radical and republican theories on the real nature and ultimate meaning of democracy are almost wholly false. The existence of certain tendencies may be an undoubted fact, yet the earnest and sympathetic observer may none the less feel compelled to deny emphatically that the ultimate outcome of these tendencies is to authorize government by the people in the sense usually ascribed to that expression. There are, in fact, two democratic demands or anticipations at the present day, which though usually identified are yet perfectly different. The first of these demands, that the people should possess sufficient influence to compel the earnest attention of government to the welfare of even the lowest in the land, is based upon a true view of political development. This, as we have already seen, is the essential meaning of representation, which modifies the constitution of Parliament until it gradually becomes qualified by the universality of its sympathies to be the protector of the whole people without distinction of class. The second of these beliefs, that the new-born political power of the people either authorizes them as a body to conduct the administration of the country, or gives them the ability to do so, is an assumption emphatically contradicted by the whole previous course of the evolution of government.

To take the first of these positions. It is frequently urged in condemnation of the democratic movement, that as soon as ever the labouring classes had gained a hold upon the political situation, they immediately proceeded to utilize their position for the selfish purpose of gaining advantages for themselves. No proceeding could in reality be more proper or natural, more in harmony, that is, with the historical process by which the sphere of parliamentary sympathies has been enlarged, and government has been compelled successively to perform its duties first to one and then to another of the different classes of the nation. When the labouring class insist that special attention shall now be directed to their own case, they are doing no more than the middle class and the aristocratic class did before them in a similar situation, when they in turn gained controlling power. The interests neither of the aristocracy nor of the middle classes enjoyed the recognition of government until they were enabled to enforce it. The influence which democracy has wielded in this respect, if properly and fairly regarded, has unquestionably implied the welfare not merely of themselves, but of the whole of the rest of the nation. It is a matter for national congratulation that, in consequence of the extension of the franchise, the influence of the wage-earners has secured recognition of the fact that care for the social and economic welfare of the very lowest of the people is a part of the duty of government. In this respect the power that has passed to the people is being utilized in a perfectly legitimate way, and for a purpose which has the sanction of the highest instincts of the race. When, however, we come to the second demand, that government should not only be administered in the interests of the people but that the people themselves should take part in its actual administration, we are dealing with a proposition of a very different kind. It is at this point that the admissible pretensions of democracy must be sharply distinguished from assumptions that have no sanction in evolution or warrant in reason. Power, we are incessantly told, has in modern times passed to the people. But this view is incomplete and misleading without more precise definition. The power of control, previously exercised first by the aristocracy, then by the mercantile classes, has been extended so as to include the labouring classes. But the wage-earners are in consequence no more entitled to call themselves the government than were the aristocracy or the middle class before them. They have not received the right nor do they possess the capacity to assume the entire conduct of the affairs of the community. The general power of control has not justified the subordination of the interests of the upper classes entirely to the lower, still less has it conferred upon them the talent required to solve the new and unforeseen problems of modern imperial administration. In what, then, does this vaunted power of the people consist? Is it power to call into existence the supreme abilities necessary to the leaders of any State that hopes to rise triumphant over all the internal and external perplexities of its own complicated existence? Is it power to create at will a Themistocles, a Pitt, a Bismarck ? The vote of the people may reject a great man, but it cannot create one. Is it power to guide the State through the intricacies of foreign policy, or to decide in a collective capacity on the measures necessary at the time of some great domestic or international crisis ? That is the sort of power democracy never has possessed, and never will. The individual genius which is indispensable for the solution of great political problems can never be satisfactorily replaced by the lucubrations of a delegated council. When we speak of the disinherited democracy coming to its own, that does not imply the ability to exercise the actual functions of government, but merely the right to share the advantages of a governmental solicitude and consideration from which it has hitherto unfortunately been excluded. Government belongs of right not to a crowd, or even to a parliamentary assembly—politeness compels us to make the distinction—but to individual minds, and to the best of those. This truth at least aristocratic and middle-class rulers have generally recognized, and have to that extent complied with the most essential requirements of national greatness. It has been left for democracy to repeat the vice of absolute kings, and to attempt to rob the nation of the services of its greatest men.

This danger is not visionary. Not only is democracy making this attempt to govern, but from the tone adopted by its champions we might infer that it was on the high road to success. The method by which they hope to attain their object is well known. The representatives of the labouring classes in Parliament are to be increased in number until they form a party sufficiently strong to control "the machinery of government." In this scheme there is no conception of government as a living vital force, employing exceptional ability on an exceptionally difficult task. It is merely regarded as a dull, dead mechanism, set in motion by representatives of the labouring classes, who alone constitute the nation. The very term "machinery of government" of which we hear so much, and the idea that if the labouring classes can "get hold of this machinery" they can secure the realization of the rest of their programme as a matter of course, implies a complete misapprehension of the fundamental nature of the problem. The conception of government as a piece of machinery implies that it is worked not by statesmen but by automatons, mere agents devoid of the individuality and initiative which characterize real government. Government by delegation implies the suppression of individual initiative, and thus in the act of being realized cuts off the vital current which forms the driving power of an efficient administration. The vote of the people may, indeed, as we have seen, confer upon their delegates an unreal and delusive supremacy; but if at the same time it negatives the presence of those characteristics by which alone supremacy is justified, and which alone are competent to deal with the complexities of the political situation, it eliminates the most indispensable element of a living, active, and intelligent administration.

But the elective system is liable to miscarry for other reasons than the selfishness, the conceit, or the ignorance of those who nominate the leaders of the State. It may be partly conscious of the value of wise statesmanship, and may yet fail through the careless application of the wrong tests of political greatness. The qualification which is most in favour at the present day, chiefly in consequence of democratic ideals, is eloquence. To such an extent does eloquence tend to take the place of the more sterling qualities required by public life that the influence of oratory may be simply but not too forcibly described as "the curse of the tongue." It is not intended to recapitulate all the well-known evils which the demagogue and the mob orator may bring upon a nation by his skill in swaying the multitude for personal or party ends. Of the mischief that may be attempted in this way by inflaming class hatred we have indeed had many recent instances. But with the education of the democracy the dangers of eloquence have assumed a new form. It is not so much that the people allow themselves to be wheedled by oratory, as that they seem deliberately to invest it with qualities which have no foundation in fact. Since the famous Midlothian campaign, few English statesmen can be accused of revelling in a profuse display of verbosity, merely to enjoy their sense of power over the audience: it is the people that are most deeply to blame for persistently regarding a mere capacity for eloquence as the equivalent of political greatness. The result is that a hopeless confusion arises as to what constitutes real merit in a statesman. The beauty of a speech which introduces or defends a policy is apparently of more account in the eyes of the British democracy than the policy itself; a little high-flown rhetoric will atone for the most glaring want of statesmanlike foresight, and a few hours' eloquence are apparently sufficient to obliterate the traces left by years of scandalous neglect. The test of administrative efficiency in English politics at the present day is in danger of being reduced to one qualification, the power of the tongue. Government by speech-makers tends to take the place of government by men of action. Eloquence, however, does not in itself promote either the internal or the external welfare of a nation, it adds nothing whatever to the real greatness of a people, and the tendency to regard it as an end in itself is irreconcilable with a sound and clear view as to what are the conditions of national welfare, and the qualities in statesmen by which this object may best be attained. It is not a little remarkable in this connexion that some of our greatest administrators have been and are men who have escaped the distorting influence of parliamentary ideals. If oratory, or at least the power of expression, is a necessary quality in a statesman, especially under democratic conditions, what is required is nothing more than the simple and direct eloquence of a Cæsar or a Bismarck. It has been suggested in a previous chapter that some of the features that made the Romans great have been reproduced under English political conditions. If the comparison is to be thoroughly deserved, it is well to remark that though the Romans regarded rhetoric as a most important part of education, yet they seem to have instinctively avoided the mistake of identifying it with statesmanship: it was kept in its place and regarded merely as an aid to political power, not as the main qualification for it. In their political literature the antithesis between words and deeds recurs with an almost irritating frequency. Yet it shows their common sense if by emphasizing this point they were enabled to escape a dangerous delusion which has become an obsession of English political life. A profuse facility of utterance is not necessarily an indication of a shallow mind, but it is instructive to observe how frequently it is associated with shallow statesmanship.

Oratory, however, curiously enough, may have uses which are unknown to most writers on political philosophy: and if it does not directly contribute to the greatness of a nation, it is at least useful under present conditions in promoting the success of rising statesmen. Mr Godkin takes the view that the chance of earning distinction in public parliamentary debate is a necessary means of attracting men of ability to take part in the work of government. This, in his opinion, has been one of the merits of the House of Commons from the first. "It was a place in which any member, however humble his beginning, had a chance to make fame as an orator. In recent days legislatures in all the democratic countries have been made repulsive to men of mark by the pains taken to get business done, and to keep down the flood of speech. Everybody who enters a legislature now for the first time, especially if he is a man of talent and character, is bitterly disappointed by finding that the rules take from him nearly every opportunity of distinction, and in addition condemn him to a great deal of obscure drudgery."[1] If Mr Godkin is right in holding that the bait of rhetorical fame is necessary to attract men of talent to the service of government, a belief in absolute agreement with the fundamental assumption of the present work, it is obvious that opportunities for oratorical distinction should not be unduly curtailed. Disraeli, we may remember, regarded the success of a certain speech as the turning-point in his career, and as the decisive influence which established him firmly on his triumphal upward progress. But if we admit the cogency of this argument, it is equally obvious that we must at the same time be on our guard against the dangers which such a policy may entail. It may be necessary to allow politicians

[1] Godkin's "Democracy," pp. 115, 116.

in this way to attract the attention of the public, but it should at the same time be clearly understood that such displays are merely of the nature of a personal advertisement, and do not in the least degree enhance the real value of a statesman to his country.

This strange inability to perceive that eloquence belongs to a merely secondary position in the rank of political excellences is largely the result of the existence of the party system. In this connexion the dangerously misleading nature of the oratorical qualifications now so much in vogue will presently appear in a still more convincing light. One of the worst defects of the party system, as already pointed out in the previous chapter, is that it prevents the utilization in government of the best political material at the disposal of the nation. It achieves this most undesirable result by the suppression of political individuality, by the undue encouragement of eloquence, and by the utterly false insinuation that the qualities most useful in the leader of a party are also most useful in the leader of a State. Of these several defects what we may call the deprivation of the right of individuality is perhaps the most severely to be condemned. It is an obvious implication of the present theory that the ablest men should not only be encouraged to rise to the highest governmental position, but should also be permitted freely to exercise their individual ability to the utmost national advantage. In order to attain this ideal a certain independence of thought and character is absolutely imperative. With the instinct of profound statesmanship Burke realized that the efficiency of government, even under representative conditions, depended, so to speak, on the self-realization of the statesman, and endeavoured unsuccessfully to obtain a recognition of this truth from the electors of Bristol. The education of democracy in this respect has not proceeded much further since Burke threw down the gauntlet to his constituents in defence of a great principle. One of the worst features of the party system in the hands of democracy is that it seeks to fetter this necessary liberty of action by imposing its views and ideals absolutely upon its representatives. Political thinking is now done not individually but in platoons. It should be a principle in politics as elsewhere that discipline ought not to be carried to a point where it is destructive of all individual initiative. The party system, in the endeavour to secure complete unanimity in the face of the common foe, tends to prevent the development of that individuality upon which the character of future leadership largely depends.

But if union in the face of the party foe is the first consideration, capacity to inflict injury on the hated enemy is the second, and this supplies the main reason for the unhealthy premium which is placed on the power of the tongue. Incisive utterance becomes an indispensable qualification for political success, and this false standard of excellence is thus imposed upon the whole nation. The result is that there is a keen demand for those merely rhetorical capabilities which are most useful to the party, but which are no necessary accompaniment of the political prescience and genius for administration which characterize the true statesman. A talent for coining electioneering phrases and clever political repartees is of more value to a party politician in his upward career than fundamental qualities of character and judgment. The nation is, in fact, bound hand and foot by a system which makes party service and platform brilliance the first requisite and statesmanlike ability only the second. As a significant illustration of this truth we may take a leading article in the *Fortnightly Review*,[1] evidently from an inspired writer. We may there discern what in the opinion of an average, perhaps a leading, political thinker, are the most desirable qualifications for the government of England. The characteristics which would be most useful in possible members of the then forthcoming administration are thus foreshadowed. " Brilliant debater and very popular speaker." "The Prime Minister should be able to address great popular audiences." " Public meetings are part of the business of politics." " Oratorical weakness might be overlooked " under certain contingencies. " Readiest, wittiest and most dashing parliamentary swordsman." The writer does incidentally refer to the " Higher

[1] March 1905.

qualities which are required for the work of government," but seems to treat them as a quite subordinate consideration. Meanwhile a Prime Minister so chosen is occupied chiefly in schemes with which the national interests are at best remotely identified, increasing his majority, conciliating opposed interests, preventing the desertion of followers. His engagements as leader of a party leave him little time for his duties as leader of the nation, even if they do not force him to pursue or defend conduct which is nationally disastrous.

Such are the types demanded by the party system; such are the paltry accomplishments required in the leader of a mighty empire. The original difficulty of distinguishing real political capacity is enormously increased as a consequence of the demand for spurious party qualifications. Superiority in a contest conducted on principles of this kind implies the survival not of the fittest statesmen, but of subtle lawyers and clever phrasemakers, types admirably adapted to succeed under the present unhealthy conditions, but which are proportionately unfitted for the honest conduct of the State. Surely it is not too much to ask that the qualities by which men rise to the exercise of imperial power should bear a closer relation to the qualities which alone can make an empire great.

Finally, the party system is undermining the foundations of the constitution and offending against the first principles of political efficiency in a way which is almost equally pernicious, but which is almost entirely ignored. It has been laid down as a postulate throughout this work that the best results of government by ability can be realized only in connexion with a strong parliamentary control, and that one of the chief functions of a political assembly is "to punish the misuse of powers entrusted to leading statesmen." But the party system has ruined the efficiency of Parliament in this respect, and has robbed it of that very function which is in reality one of the most essential features of an assembly which exists for the very purpose of protecting the widest national interests of the community. Under the party system the criticizing and amending powers of Parliament are ironically conceded to a practically helpless minority, the opposition; while so far is the protective organ of the nation from being ready to punish the misuse of political power that its main endeavours are under present conditions frequently concentrated upon the defence of conduct which has been proved to be politically disastrous or even personally dishonourable, and which in one case at least has inflicted irreparable injury on the nation. The disorganization of government, notorious at the present day, is not due as many suppose to the excessive authority of the Cabinet, but to the annihilation of the controlling power of Parliament, a very different thing. Ministers in appearance act tyrannically, but they are in reality puppets, helplessly impelled, at least on the Liberal side, by a party which goads them on from behind. The function of a Government is to govern, and not to be hopelessly impelled from behind. Until the restrictive functions of Parliament are restored we shall enjoy, not the carefully controlled government of intelligent statesmen, but the worst possible form of administration known in history, government by faction. One of the most obvious criticisms on the present theory is doubtless to the effect that if consistently carried into action, it would tend unduly to encourage the dangers of personal government. This objection will be examined later on, but it may at once be stated that it comes with the least possible weight and the worst possible grace from any who acquiesce in an habitual misuse of the powers of Parliament, such as is prevalent at the present day. The authority of the national assembly is deliberately invoked, under the present system, for the avowed purpose of protecting ministers who have brought the nation within measurable distance of irremediable disaster. Perhaps the most unpardonable vice of which pretentious statesmen can be guilty is to mistake the nature of the world in which they live, and those who endanger the community by such miscalculations should meet with appropriate condemnation. But beneath the motherly care of the parliamentary party system ministers may confidently count on escaping all practical inconvenience for indulging in the most fatal delusions, and may look forward to unabashed resumption of duties for which they have proved themselves politically unfitted. To correct so grave an abuse of constitutional powers, and to make the House of Commons a freely controlling and amending body, instead of the mere accomplice of political malefactors, and the unworthy instrument of hasty schemes of dishonest and even treacherous legislation, would seem to be an object worthy of the united efforts of all who realize how fatally the fundamental principle of effective criticism has in recent years been impaired. Yet the difficulties in the way of such a task are apparent, when we reflect that the real reason why the House of Lords has been marked down for mutilation or destruction is not because it possesses a somewhat antiquated constitution, but because its retention of the power of unrestricted comment, and its disinterested attachment to certain axioms of sound and healthy government, is offensive to the autocratic instincts of democracy.

But if the democratic encouragement of the party system has robbed Parliament of its proper controlling functions, and by imposing a false standard of excellence has also rendered difficult the emergence of true imperial greatness, England is at least free from a certain grave weakness of other democratic nations. Democracy in other countries is opposed to the principle of government by the fittest in consequence of its jealousy of political talent, and its careful cultivation of the belief that one man is as good as another for the purpose of national administration. In this principle, which strikes at the very root of the progressive development of the State, is to be found the true explanation of the unsatisfactory nature of American political conditions. Political supremacy is an offence to a community theoretically based on the assumption that all men are equal. The consequence is that in the first place the Government of America is deprived of the assistance of the great abilities which have had so marked an influence upon industry and commerce, and which under more favourable conditions would seek the ranks of the administration. In the second place, as in the case of the French under Louis XIV., the individual genius and the desire for personal distinction which have been diverted from their proper occupation and natural outlet in politics, have begun to assert a form of supremacy which is prejudicial to the interests of the community. We have seen how the exclusion of the French nobility from the work of government weakened the intellectual character of the administration, while their baffled ambition turned in a direction which was ruinous to the welfare of the State. A study of American political life likewise reveals the fact that the discouragement of the higher kind of talent from asserting its natural right to direct the government of the country, has resulted partly in a lowering of the character of politics, and partly in an inordinate pursuit of commercial success. The consequences which have ensued from the lowered tone of politics are well known. It is not so generally recognized that the worship of the dollar is at once a source of moral weakness to the nation and of political instability to the State.

It is very generally asserted that the best men in America decline to enter political life because of its low and corrupt tone. But this is an inverted method of stating the problem. If we are to trace the phenomenon to its real cause, we must put the case the other way about. It is because the best men have been deliberately prevented by democratic theory from assuming their rightful position in the State that corruption in politics has ensued. Those unattractive features in political life which now cause the voluntary abstention of the highest talent from participation in the work of government are themselves the result of the enforced exclusion of the highest talent from the work of government in the past. The indulgence of the deep-seated instinct which urges men to seek for an honourable pre-eminence in politics, and which would have led to the domination of the best, has been prevented only to lead to the domination of the worst, or at least to the influence of a political authority of an undeniably disreputable character. The attempt to withhold any acknowledgment of the natural superiority of great individual leaders in the higher ranks of government has produced the unnatural supremacy of the "boss" in a lower but most vital department of the political system, who has reached out his hand to grasp that political ascendancy from which better men have been warned off. The decreasing power of the President and the increasing power of the speaker of the House of Representatives appears to be due to the same cause. The speaker has a power, says Bryce, "which in the hands of a capable and ambitious man becomes so far-reaching that it is no exaggeration to call him the second if not the first political figure in the United States, with an influence upon the fortunes of men and the course of domestic events superior, in ordinary times, to the President's."[1] The increased authority of the speaker may not be as pernicious as that of the boss, but it equally shows that the ill-judged effort to prevent motives of ambition from affecting the appointment to the highest governmental office, results in an inconvenient weakening of presidential authority, and in the transference of political power to a quite unexpected and much less desirable quarter. In defiance of the constitutional arrangement which attempted to render for ever impossible any undue personal influence over the "legislature," the natural rivalry of individuals in a political assembly triumphantly asserts itself, and in Congress as in Parliament, the strongest character comes to the front, and assumes a power not contemplated by

[1] Bryce's "American Commonwealth," vol. i. p. 140.

the founders of the State. And it is for such results as these that the Americans have deliberately deprived themselves of the talented political guidance unquestionably at their disposal. "Since the departure from the stage of the old political worthies," wrote W. R. Greg more than fifty years ago, "America has had three statesmen of high capacity and European reputation—Clay, Calhoun, and Webster; and all have aspired to the Presidency in vain." Another fifty years have elapsed, and the democratic system has since then successfully prevented the appearance of real political talent, not only in the Presidency but anywhere else; and as a natural consequence politics, which seemed to the Romans the noblest of all occupations, a trial and an enterprise worthy of the highest powers of man, has sunk to the level of a second-class profession, or has even come to be regarded as one of the less reputable methods of earning one's daily bread. And strangely enough, it was the very depth of the respect entertained for the duties of government by the founders of the American Constitution which led them, in their attempt to purify politics, into defiance of the laws of nature; and this in turn has produced a condition of things so remote from their original aim, that instead of forming a model to more backward European communities, their political system has become an acknowledged scandal in the eyes of the civilized world.

But this virtual prohibition of the political

career in a nation containing a more than ordinary proportion of able, active, and ambitious men, has, in addition to lowering the intellectual character of the administration, another unfortunate effect of an almost equally unwholesome kind. Excluded by democratic prejudice from its proper domain of political activity, the superior talent of the individual has turned to the economic sphere, and has been driven to the amassing of large fortunes in order to satisfy the ineradicable love of power. The enormous production of wealth in America is generally regarded merely as triumphant evidence of the economic efficiency of a community when freed from the senseless restrictions which hamper feudal and aristocratic Europe. But this is a too flattering representation of the matter. It is hardly too much to say that this inordinate acquisition of riches, being the result of the too exclusive attention of great men to merely pecuniary ambitions, is a sign of disease and not of health in the body politic where it occurs. If from one point of view this phenomenon affords evidence of widespread individual energy, and may so far be regarded as satisfactory, from another point of view it shows with equal certainty that a serious misdirection of individual talent is taking place. The ability which should have been devoted to the art of government has been diverted to the creation of large fortunes, which, so far from assisting the general welfare, properly so regarded, have become a source of menace and danger to the social and even the political progress of the community. Thus democratic theory, by endeavouring to avoid the imaginary danger of the supremacy of the individual in politics, has ended by producing or helping to produce a supremacy of individuals in the economic sphere, which constitutes a real danger to the healthy political development of the State. The Americans insistently demand the admiration of the world on the ground that they have had the courage to rid themselves of the prejudicial interference of kings and nobles in their political affairs. But a republic which prides itself upon having abolished monarchy and aristocracy merely to make room for the super-millionaire, and which seems to countenance the claims of real political genius as little as it tolerates those of empty aristocratic rank, is in danger of meriting a severity of criticism similar to that which it bestows upon "the effete systems of Europe."

An examination, then, of the democratic tendencies both of the present and of the past leads unmistakably to the conclusion that democracy is guilty of an offence for which nature as revealed in history has no forgiveness, the offence of preventing the ablest men from reaching the highest positions in the State. The qualifications for political supremacy in a democratic State are for the most part not the qualifications for really able government: and even if really able men do reach a position of influence, they cannot use their ability in the most efficient way, because their efforts are hampered and distracted by unsound theories of political action for which democratic tendencies are mainly responsible. It is democracy that now claims the right divine to govern wrong, and the liberty of appointing ministers at their own will and pleasure, irrespective of the highest interests of the State. Democracy, like autocracy, has, in fact, put forward pretensions which it could not justify and has attempted a task which it could not perform. It will therefore suffer the fate which in England has always overtaken incompetent rulers in the past. Englishmen accustomed for centuries to a real though unacknowledged system of government by natural selection will definitely revolt against a theory of government of which the logical outcome is to degrade statesmen to the rank of trade union officials. The path will presently be cleared for the unfettered activity of real political talent, and real tests of political efficiency will take the place of the spurious qualifications consecrated by democracy and the party system. This counter-revolution will, in fact, undertake the destruction, not of aristocratic but of democratic privilege, which claims the right to fill the offices of State with the subservient instruments of democratic policy. The whole of the real intellect of the country on the one side will eventually be found arrayed against mere weight and numbers on the other. As a matter of fact this revolution is already in progress, and there are not wanting indications that the country is seriously beginning to inquire into the title and qualifications of democracy to exercise this supreme administration of affairs, and to find the title baseless and the qualifications non-existent. Aristocratic privilege lasted long because it justified itself by real national service. The days of democratic privilege are already numbered, because, though capable of fighting on its own behalf, democracy has shown itself incapable of securing the wider interests of the nation as a whole. The ruin of democracy in the past and its failure in the present is attributable to one fundamental cause—that the people have mistaken the nature of their functions in political evolution. They are aware that the progress of political development has, by universal consent, placed power in their hands, but they have mistaken the nature of that power. They have been assured that all other classes must give way before them, that the divine right of the people has replaced the divine right of kings. But the legacy of kingly power has not implied the gift of statesmanship. If past history is any guide as to true and false methods in politics, the success of nations is inseparable from talented individual administration. The work which statesmen have performed in the past, statesmen must perform still, if, at least, national health is the real consideration. It is the part of the people to strengthen the influence of great statesmen by their authority and to encourage them by their commendation, but not to usurp their place. It is their function to lend their collective power to support great rulers in their high position, but not to rule themselves. Those excellences of the constitution in which England has long been acknowledged to lead the world do not consist either in liberty alone or in parliamentary institutions alone, but in a liberty which encourages able men to assume political leadership under the advice and control of Parliament. The English constitution is great, not because it is an elaborate system of checks and balances, but because it combines the intellectual advantages of government by ability wherever found with the moral advantages of parliamentary watchfulness. The highest function of democracy is to increase the intelligence and the honesty of the all-watchful care of the controlling power of Parliament.

A theory which lays great stress upon ability in government will perhaps be considered open to the objection that mere intellect does not necessarily imply the presence of those qualities which in a ruler are best calculated to ensure the happiness and welfare of the State. But nowhere in these pages has mere intellectual brilliance been assigned a greater importance than personality and character. Should it be urged, however, that there is an occasional uncertainty about the action of genius in politics, and that there have been periods in history when particular nations would have fared better under men of lesser brilliance and sounder political views, the answer to this objection is obvious. In the first place, this is a risk which must be taken, and in England at least it may be taken without serious misgiving, since the political instinct of the nation has always been sufficiently healthy to prevent the endurance of unsound statesmanship for more than a brief period. In the second place, if a sufficiently extensive survey be taken of history, this difficulty will be found to disappear. Even if it be admitted that one clever man may sometimes make a mistake, yet a succession of able men, such as the free course of political selection would produce, necessarily tends to correct occasional aberration in matters of State. It is to this continuous succession of relatively able men, rather than to the presence of distinguished genius, that the superiority of administration in Roman and English history must be attributed. If any of the communities already celebrated in history had from its earliest beginnings been enabled to carry out consistently a perfectly natural selection of its statesmen, it seems beyond question that it would have reached a condition of prosperity far exceeding that to which it did actually attain.

Those devotees of freedom who disguise their jealousy of other people's greatness under the dignified appearance of a protest against personal government, and who affect to believe that so long as certain theories and principles of politics are accepted it is a matter of perfect indifference who is chosen to carry them into operation, will doubtless find the main contention of the present work in the highest degree unacceptable. The purity of the motive which inspires their objection, not above suspicion in its origin, is rendered more than ever doubtful by the fact that it is advanced mainly by those who have degraded Parliament to the level of the servile instrument of a dangerous political coalition. In the next place, the conviction is growing in many minds at the present day, that the incessant advertisement of infallible programmes and policies is not the natural symptom of a sound political development, but the entirely artificial result of the unhealthy predominance of the party system. It is a pleasant fiction of theorists that a party exists to promote certain lines of policy. It would be much nearer the mark to say that policies are deliberately invented, and differences of opinion unnaturally created, in order to justify, or at least to excuse, the previous existence of the party. The consequence is that the correction of a mistaken policy is rendered more difficult than ever because the party system makes it a personal and a political humiliation to accept even the correct views of an opponent. If, again, the elevation of some theory like that of free trade, which has a strictly relative and temporary expediency, to the rank of an absolute and eternal truth, is a fair instance of the tendency of government by policies, the less we have of such *a priori* and doctrinaire methods the better. When we add that a dishonest and vote-catching frame of mind is deliberately encouraged by the system of "programmes," we begin to feel that the concession of a greater liberty to honest individual influences and convictions would be a welcome and a wholesome change. And even granting the value of correct theory, the obnoxious element of personality is not even then eliminated. Political theories do not spring to life unbidden from some mysterious region of thought. They do not descend self-created from the political firmament entirely independent of human agency. If the policy is old, it is simply a tradition of government resulting from the political experience and discernment of certain able individuals, modified, perhaps, to suit existing conditions. If the policy is new, it is the enforcement of the decisions of leading statesmen or influential thinkers of the day. The personal element is always to be found somewhere, first and last. Great policies originate in the intellects of great men. The utmost that can be urged against the validity of the argument of the present work from this point of view is that the better man may occasionally be found supporting the inferior theory, and this objection has been already answered. Let a nation consistently throughout its history determine to take care of "the men," applying the same principle of careful selection to members of Parliament as to the members of the Cabinet, and the "measures" will take care of themselves.

In conclusion it may be not without interest to

examine the bearing of the present theory upon the relative condition of the leading nations of to-day. With regard to the international situation, the advantage that England derives from the system of government by natural selection will be at once apparent. Of all the nations of the world, England and, in a lesser degree, her colonies alone possess a political constitution, a national training, and, in consequence, an individual temperament, which facilitates the access of their ablest men to office. One reservation must, however, be made—the degrading requirements of the party system encourage spurious qualifications for office, and deny a sphere of proper activity to the really great imperial statesmen we possess. The political history of France since the times of the Directory has been such that the nation dares not encourage political pre-eminence, and disastrous parliamentary incoherence is the natural result. In America the assertion of a personal superiority in politics is regarded as an offence against the principle of individual equality, and talent is warned to take its ambitious hopes elsewhere. In Germany the dismissal of Bismarck has entailed consequences identical with those which followed the enforced resignation of the elder Pitt; ministers are not chosen according to merit; their selection, together with the entire direction of the policy of the State, is monopolized by the inferior abilities of an hereditary king. Some catastrophe may therefore be expected similar to that brought upon

England by the administration of George III. Who wields the real power in Russia it would be difficult to say; but this at least is certain that government by ability irrespective of any other consideration is far from being realized. In Canada the career for political talent is practically free, and the national and imperial influence of her great men has been proportionately beneficial. In Australia and New Zealand the conditions have also permitted and indeed encouraged the appearance of great statesmen: but it must at the same time be acknowledged that the independence of individual judgment and administration was at one time seriously impeded by the growth of ultra-democratic and socialistic ideals, which took no account of the international situation. One of the least desirable results was a determined attempt to restrict immigration, and at the very time when it was most needed. The consequence was a reduction in the strength of garrisons necessary to defend possessions which, in an over-populated world, must one day become the prize of the strongest. The pressure of stern and unpalatable fact has, however, already begun to correct the dangerous idealism of the political lotus-eaters of the Pacific. In the most essential feature of political well-being, therefore, the British Empire enjoys immense advantages, and if the encroachments of democratic privilege upon the efficiency of government are successfully restrained, parent kingdom and loyal children may be expected to continue indefinitely their widespread and beneficent activities.

To the general estimate of the value of government given at the end of the second chapter a final word may now be added. The end of government, said Locke, is the good of mankind. The end of government, said the school of Bentham, is the greatest happiness of the greatest number. Both these views express a general truth and constitute an advance upon the opinion of the times in which they were written, but they lack precision. The final end of government, we may admit, is doubtless the promotion of the highest moral and social welfare of the community. Unfortunately no nation is free to bestow the whole of its efforts upon the attainment of this end. If it is to survive in order to carry this great purpose into effect, it must be simultaneously equipped for playing a successful part in the rivalry, economic as well as military, which is the law of international life. The attainment and maintenance of the highest social welfare is conditional upon the attainment and maintenance of an equally high standard of economic and military proficiency, necessitated by the rivalry of nations in commerce and in war. The end of government, conditioned as a nation is by the laws of international rivalry, is therefore from this point of view not the greatest happiness of the greatest number, but the greatest sum of national energy of which the people is capable. The most important practical truth which universal history teaches, a truth habitually ignored by democracies, is this, that the international situation dominates the internal. The pursuit even of great ideals, as well as the accumulation of the means of material enjoyment, must be subordinate to the attainment by a community of national efficiency; and its national efficiency in turn depends upon the quality of its administration. The exacting demands of the international situation place a supreme value upon wisdom of government. Only a nation whose political system habitually encourages government by the ablest can ensure the leadership, the organization, and the wise treatment of internal and external difficulties, necessary to maintain a proud position through centuries.

INDEX

INDEX

INDEX

TURNBULL AND SPEARS, PRINTERS, EDINBURGH

For Product Safety Concerns and Information please contact our EU representative GPSR@taylorandfrancis.com Taylor & Francis Verlag GmbH, Kaufingerstraße 24, 80331 München, Germany

Batch number: 10725863

Printed by Printforce, the Netherlands